D1824320

ISBN 978-0-260-43869-0
PIBN 10279145

FB &c Ltd, Dalton House, 60 Windsor Avenue, London, SW19 2RR.
Company number 08720141. Registered in England and Wales.

For support please visit www.forgottenbooks.com

PUBLICATIONS

OF THE

Earthquake Investigation Committee

IN

FOREIGN LANGUAGES.

No. 26.

TOKYO, 1908.

PUBLICATIONS

OF THE

Earthquake Investigation Committee

IN

FOREIGN LANGUAGES.

No. 26.

TOKYO, 1908.

AN INVESTIGATION

ON

THE SECONDARY UNDULATIONS

OF

OCEANIC TIDES

CARRIED OUT BY THE ORDER OF

THE EARTHQUAKE INVESTIGATION COMMITTEE

DURING

1903-1906.

No. 1.

No. 2.

No. 3.

No. 4.

No. 1.

No. 2.

No. 3.

No. 4.

that stationary waves may be produced between the Kuroshiwo and the shore, but when the water here is once disturbed, we may expect most promiscuous groups of waves to be propagated from the disturbance; these waves will be swallowed by the bays and estuaries lying in their way, especially when the periods of oscillation are in close agreement.

Sir G. H. Darwin,* in his wellknown lectures on "Tides" delivered in 1897 speaks of the tides in Venice:—

"Every visitor to Venice must, however, have seen, or may we say smelt, the tides, which at springs have a range of some four feet. The considerable range of tide at Venice appears to indicate that the Adriatic acts as a resonator for the tidal oscillation, in the same way that a hollow vessel tuned to a particular note, picks out and resonates loudly when that note is sounded."

Later investigation in 1904, 1905 and 1906, which were mostly undertaken by Drs. Honda and Terada, assisted by several graduates of the university, in nearly all the bays bordering on the Pacific Ocean and the Japan sea, revealed the truth of the acoustical analogy first propounded by the great authority on the tides.

Led by this consideration, Dr. Terada treated the problem of the secondary oscillations by bringing in the theory of resonators in close contact with the vibration of bays and estuaries. As an outcome of the discussion, a mouth correction must be added to the period calculated according to Merian's formula, as is wellknown in the theory of organ pipes. Still more interesting is the mutual influence of the dumb-bell shaped bays, bearing close resemblance to the acoustical resonators

* Tides, p. 168–169, First Edition.

communicating through a narrow channel. This is beautifully illustrated in the oscillations of some bays, fulfilling the stated condition.

When the volume of water in the bay and consequently the depth and the breadth change, it is necessary to add a small correction. This question was discussed by Dr. Isitani.

In limnological work, Sarasin's limnimeter is universally recognised as a trustworthy and convenient instrument, and was used in some of our lake surveys. In observations of secondary oscillations, the large range of the tidal fluctuation prevented the use of the instrument. The tide gauge in its usual form is too cumbrous as a portable instrument, and the great damping through communicating tubes annihilates waves of secondary oscillations of short periods, so that they are mostly lost on the record. This defect was first modified by Dr. Nakamura by balancing the pressure by means of a mercury column, and greatly reducing the range of the tidal fluctuation without in the least interfering with damping, which it was necessary to keep suitably small in order that the secondary oscillations may not be lost to view. This was further improved by Dr. Honda, who greatly simplified the apparatus, and changed it to a neat portable form. In most of the present investigations, Dr. Honda's instrument was used.

The records obtained either by the tide gauge or with Dr. Honda's instrument all present semidiurnal fluctuations, and on it the undulations to be investigated appear in serrated form. For the exact study of the phenomena, it is sometimes necessary to eliminate the tidal undulation. This was effected by means of Dr. Terada's tidal rectifier, by which only the secondary oscillations were brought to view. When the bay responds to waves of different periods, so that they are mingled together, this procedure was the easiest step for analysing the different components.

Dr. Endrös, in his elegant research on the seiches in one of the Bavarian lakes, made use of a model for studying the periods of oscillation. The same method was followed by Drs. Honda, Terada, and Isitani in a slightly different manner. The models of bays having the contour lines and the magnified depths of those already studied were placed in a water tank, so that the water in the model came to the level mark, and waves were then excited in the tank. When the period of the wave was in harmony with that of the model bay, the water in the model responded to the exciting wave with extreme ease, and continued vibrating for some time even after the subsidence of the exciting wave. Not only was Dr. Endrös's experiment extended in this direction, but the courses of the stream lines in the model were closely studied by Dr. Honda. By an ingenious device of sprinkling the surface of water with fine aluminium powder, and photographing the surface by a camera with the optical axis vertical, the trace of dust particles was observed; these photographs proved distinctly that the surface of the water was oscillating and showed at a glance the mode of response to the external source of excitement. This graphical representation is more practical than that deduced from mathematical calculation, which is next to impossible on account of the variable depth and the irregular contour. Thus in delineating the oscillations proper to bays, the study with models, when a hydrographical chart can be obtained, is generally sufficient to determine the nature of oscillations and their periods. It seldom happened that the periods which can not be detected with the model were ever observed. We are now in a position to infer from the study of models how the bay oscillates without entering into the actual registration of the secondary oscillations.

The contour lines of equal depths in different bays were drawn by Dr. Isitani, who also undertook the laborious task of integrating and determining the volume of the bay; after suitably choosing the median line, he also calculated the periods. They generally agree quite closely with the observed values.

The problem which still remains unsolved is how the waves of different periods are generated in the surrounding ocean and especially on the Pacific side. They may be due to local variations of atmospheric pressure, earthquakes, and such allied causes. The following hint may also not be out of place as to the cause of these waves on the Pacific coast. By far the greater part of the destructive sea waves, which from time to time have devastated our coast, seem to have had their origin on the waves originating off the eastern coasts of Japan. The existence of a sort of standing waves with the land on one side and on the other the ocean current running nearly parallel to the shore is quite imaginable. Along the east coast of Rikuchû and the southern part of Hokkaidô, the gulf of Tosa and the adjacent districts, the course of the current is nearly parallel to the line of equal depth. In such cases, the presence of standing waves with the land and the current as the boundaries is a matter beyond dispute. The current will not behave exactly like a solid shore, but on account of its high speed, it will partake the nature of a quasi-elastic solid, making the waves rebound. It resembles a liquid jet in the ocean; it will oscillate with periods peculiar to itself; it will thereby be capable of transmitting vibration to the bounding waters; it will be set in forced vibrations by earthquakes and other causes. In fact, the behaviour of the current resembles a piece of india-rubber band, several hundred miles wide, stretched nearly parallel to the coast on the Pacific side. It therefore appears to me that the existence of Kuroshiwo is extremely favourable to the generation of exciting waves, which spread devastation along the sea coast.

Thus a portion of my proposals to the Earthquake Investigation Committee has been happily brought to a close by the indefatigable zeal of several investigators, who not only prosecuted the observations, but improved the apparatus, deduced corrections to observations, compared the observed with the calculated periods, and solved the important question as to the mode of excitement of oscillations proper to bays and estuaries. It fell to my lot only to guide the method and fix the places of observation, and it is on account of this responsibility that I have allowed myself to add these words as a preface.

Finally I must not omit to state that my best thanks are due to Dr. B. Mano, the President of the Earthquake Investigation Committee, for allotting during several years a part of the fund granted to the committee for continuing the present research, and the kindly interest with which he has watched the results.

H. NAGAOKA.

Member of the Earthquake
Investigation Committee.

Physical Institute,
May 1st, 1907.

CONTENTS.

Frontispieces

Preface

		Page.
§ 1.	Introduction	1–4
§ 2.	Tide-gauges	4–11
§ 3.	General results	11–17
§ 4.	Special result	
	I. Coasts of Hokkaidô	17–20
	II. Japan Sea coasts of Honshiu	20–23
	III. Pacific coasts of Honshiu	23–37
	IV. Pacific coasts of Shikoku	37–43
	V. Coasts of Kiushiu	43–48
	VI. Bonin Islands and Formosa	48–49
§ 5.	Experiments with models	49–57
§ 6.	Formula for calculating the periods of the undulation in bays	57–67
§ 7.	The method and the result of calculation for the period of oscillation	67–76
§ 8.	Sea waves and secondary undulations	76–104
§ 9.	Oscillation of large bays and anomaly of tides	104–110
Name list of stations		111–113

PLATES.

Marcograms	I–LXIV
Maps	LXV–LXXXVI
Photographs of models	LXXXVII–XCII
Weather charts	XCIII–XCV

Secondary Undulations of Oceanic Tides.

By

K. Honda, *Rigakuhakushi,* **T. Terada,** *Rigakushi,*

Y. Yoshida, *Rigakushi,* and **D. Isitani.** *Rigakushi.*

ERRATA.

Page 19, line 4, *insert* '1905' after the date.
" 54, " 12, *for* '2.28s' *read* '2.20s.'
" 73, " 14, " '39.0' " '36 0.'
" 103, " 14, " '28' " '4.'
Pl. XII, for the vertical scale of Inuboye, *replace* that similar to Pl. XI.
" XXV, in the vertical scale, *for* '50,' '100' and '150,' *read* '5,' '10' and '15.'

Milne* discussed the remarkable undulations of July, 1843, on the coasts of Great Britain, and ascribed their origin to the storm then prevailing at the district. Admiral Smyth† referred to a similar phenomenon at Mazzara, Sicily, where it had long been termed *Mirabia* or *Marrobbio.* Sir George Airy‡ studied the phenomenon at Malta and believed that it was due to

---

*) D. Milne, On a remarkable oscillation of the sea observed at various coasts of Great Britain, Trans. Roy. Soc. Edin., 1844.

†) Admiral Smyth, Memoir descriptive of the resources, inhabitants and hydrography of Sicily (London, 1824).

‡) G. Airy, On tides in Malta, Phil. Trans. Roy. Soc. London, **169**, 1878.

seiches between Sicily and the African coasts. The phenomenon was also observed on another coast of Italy and at the northern coasts of Europe.*

In 1895, W. Bell Dawson† communicated a paper to the Royal Society of Canada, which showed the existence of the secondary undulations of considerable amplitude. Professor Duff‡ concluded from the observations at many stations along the coasts of the Bay of Fundy and the Gulf of St. Lawrence, that the secondary undulation had a period peculiar to each station, and that it is of the nature of seiches excited by the low atmospheric pressure. H. C. Russell** states that at Sydney, the undulations are in most cases due to atmospheric disturbances. Napier Denison†† of Toronto Observatory made a systematic study of the subject in connection with the barometric change. He attributed the phenomenon to long waves generated by air waves of considerable wave length, accompanying the low barometric pressure. Since several years ago Professor Giovanni Platania*† has investigated the secondary undulations in the Gulf of Catania. He attributed the phenomenon to the

---

*) G. Groblovitz, Ricerche sulle Maree d'Ischia, Rend. Acc. Lincei, **6**, p. 26-32, 1890. Sulle osservazioni mareografiche in Italia e specialmente su quelle fatte ad Ischia, Atti del I Congr. Ital., Genova, 1893.

R. Sieger, Niveauveränderungen an Skandinavischen Seen und Küsten, Verb. 9te Deutsch. Geogr., Wien, p. 224.

E. Brückner, Ueber Schwankungen der Seen und Meere; Verb. 9th Deutsch Geogr., Wien, p. 209, 1892.

†) Dawson, Notes on secondary undulations, Proc. Roy. Soc. Canada, May. 1895.

‡) Duff, Seiches on the bay of Fundy, Amer. Journ. Sci., **3**, p. 406-412, 1897. Periodic tides, Nature, **59**, 1899.

) Russell, The source of periodic waves, Nature **57, 1898.

††) Denison, Secondary undulations of tide-gauges, Proc. Can. Inst., **1**, p. 28, 1898. The origin of tidal secondary undulations, Ibid, **1**, p. 134.

*†) Platania, Le librazioni del mare con particolare riguardo al Golfo di Catania, Atti del V Congresso Geogr. Ital., Napoli, 1904. Nuove ricerche sulle librazioni del mare, Annuario der R. Istituto Nautico, I, 1907.

plurinodal stationary oscillation of the bay, and observed that oscillations of conspicuous amplitude occur in connection with barometric disturbances.

In the bays on the Pacific coasts of Japan, the phenomenon is sometimes so conspicuous that it is commonly known as *yota*. In the harbour of Nagasaki, it is called the *abiki*; its amplitude of oscillation frequently exceeds 50 or 60 cms. The *yota* or *abiki* is frequently accompanied with calm weather preceding low barometric pressure, which is approaching our coasts.

Professor F. Omori* investigated the secondary undulations of Ayukawa, Misaki and Hososhima mareograms in connection with the discussion of the several destructive sea waves, and found that the periods of the waves are the same as those observed in ordinary cases. The records of the tide-gauges on Indian coasts†, of the waves which accompanied the great eruption of Krakatoa, 1883, were also found to show the same periods as frequently occur in ordinary cases. He explained this interesting fact by the consideration that a bay or a certain portion of the sea oscillates like a fluid pendulum with its own period. Professor H. Nagaoka,‡ in his paper on the hydrodynamical investigation of the destructive sea waves, expressed the desirability of a special inquiry concerning the phenomenon. The suggestion was taken up by the Earthquake Investigation Committee, and the task of carrying out a systematic observation was imposed upon us.

The observations were carried out during the period of 1903

*) Omori, Publications of Earthquake Investigation Committee **34**, p. 5, 1900.

†) Omori, Proc. Tôkyô. Math.-Phys. Soc. **2**, p. 455, 1905. Publications of Earth. Inves. Comm. **56**, p. 29, 1906.

‡) Nagaoka, Proc. Tôkyô Math.-Phys. Soc. **1**, p. 126, 1903.

to 1906; the number of bays and coasts observed amounts to fifty one in all.

The observation consisted firstly in finding the periods of the undulations peculiar to each bay or coast, and secondly in comparing the phases of the undulation in the different portions of a bay by simultaneous observations.

In all cases, portable self-recording tide gauges were used.

The research on the phase relation was carried out only for several typical bays.

Fig. 1.

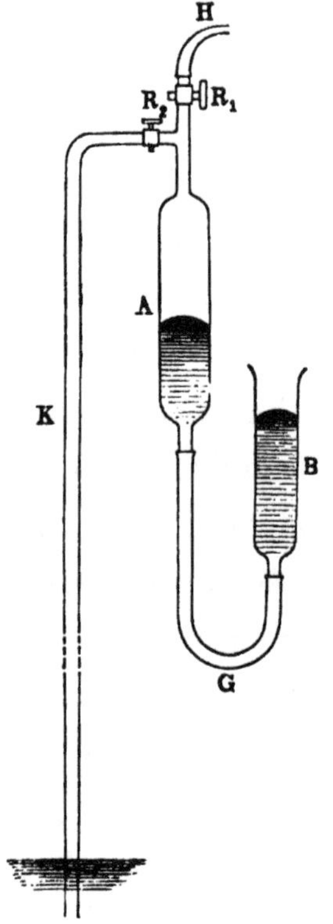

§ 2. TIDE-GAUGES.

The tide-gauge used in our first excursion of the summer vacation of 1903 was that designed by Assistant Professor S. Nakamura.*

The tide-gauge is very simple and portable; Fig. 1 shows diagramatically the principle of the apparatus. *A* and *B* are cylindrical vessels of glass; *A* is connected to sea-water by a lead tube *K* through the cock R_2. A suitable quantity of mercury is put in *A* and *B*, which communicate with each other by a thick caoutchouc tube *G*. By sucking a caoutchouc tube *H*, water is brought into the vessel *A*; when the vessel and the lead tube are

* S. Nakamura, Proc. Tôkyô Math.-Phys. Soc., 1, p. 123, 1902.

completely filled with water, the cock R_1 is closed. As the level of the sea changes, the mercury in the vessel B moves up or down; if it is so arranged that the up and down motions of the mercury are recorded on a vertical cylinder rotating with a definite rate about its vertical axis, then the tide is traced on the cylinder on a reduced scale. The cylinder (20 cm. high and 9.4 cm. in diameter) usually rotates once per day; it can also be made to rotate once in two hours. The cock R_2 serves for damping the rapid up and down motions of mercury due to the surface waves.

The recording arrangement of the tide-gauge was, in our case, modified in the following way (Fig. 2). The float made of a hollow ebonite box loosely fitted the vessel B, and on the box, a thin aluminium rod was vertically erected. A pen-holder p, which carried two arms perpendicular to the pen, was horizontally fixed to the rod. At each end of these arms, a friction wheel was fixed, which rolled between the V-shaped grooves of two vertical guides gg.

Fig. 2.

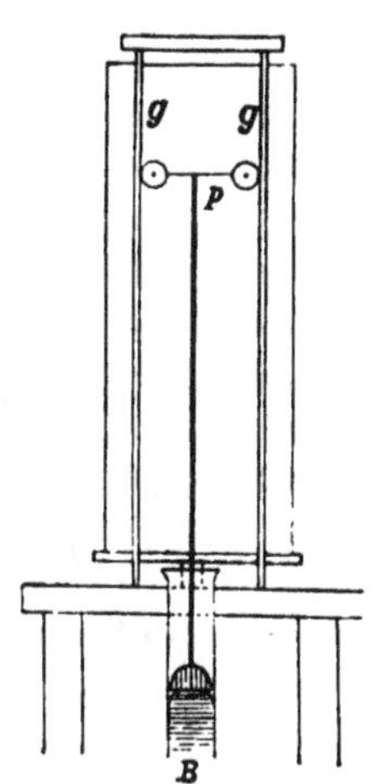

In this way, the pen was constrained to move in a vertical line; though the pen was slightly pressed against the recording cylinder E by a weak spring, its friction was quite insensible. The photographs of the whole apparatus and the recording pen are given in the frontispiece, Photo. No. 1 and 2.

If h_1, h_2 and h_3 be the heights of the levels of sea water and of mercury in A and B measured from the bottom of the sea, S_2 and S_3 be the cross sections of A and B respectively, then the pressure P of the water on the surface of mercury in A is given by

$$P = \pi - (h_2 - h_1) = \pi - \rho\,(h_2 - h_3),$$

when π is the atmospheric pressure and ρ the density of mercury.

We have also the equation of continuity,

$$S_2\,dh_2 = -S_3\,dh_3\,;$$

taking the differential of the first equation, we have

$$dh_2 - dh_1 = \rho\,(dh_2 - dh_3)\,;$$

eliminating dh_2, we get

$$\frac{dh_3}{dh_1} = \frac{1}{\rho\left(1+\frac{S_3}{S_2}\right) - \frac{S_3}{S_2}}$$

This reduction-factor was nearly $\frac{1}{18}$ in our tide-gauge. The present apparatus worked very satisfactorily, when it was possible to set the instrument near the shore and the height of the mercury in A or B above sea level did not exceed 1 or 2 meters. The practical difficulty, however, met with in our excursion was that it was often necessary to set up the apparatus at high stations, where the sucking was difficult and the air dissolved in water frequently gathered in the vessel A, and caused an unusual rise of the pen. To avoid these inconveniences, another arrangement has been used since the winter excursion of 1903. It is in fact a modification of Richard's tide-gauge; in our case, his recording arrangement was replaced by ours, and the caoutchouc bag in his diving jar was dispensed with. Our diving jar is shown in Fig. 3. A is a closed cylindrical vessel made of brass, 12 cm. high and 12 cm. in diameter; it is screwed into a heavy lead disk *D*. By the tube *a*, water enters the vessel and compresses the enclosed air. The vessel *A* communicates with the recording apparatus by means of a brass tube *b* and a long copper tubing *l* (internal diameter=2 mm.); the brass tube is bent round as shown in the above figure for convenience of transportation. The recording arrangement is almost the same as the preceeding. In Fig. 4, *B* and *C* are two glass vessels communicating with each other by a thick caoutchouc tube, and partially filled with mercury. The copper tubing *l* from the diving jar is connected to the glass vessel *B*. *p* is the pen, *gg* the guiding pillars for vertical motion of the pen, and *E* the recording cylinder. If the jar is placed on sea bottom, the pressure of the enclosed air is balanced by the pressure due to the difference of the heights of two mercury columns. The change of pressure caused by the change of sea level above the diving jar, forces the motion of the mercury column in the vessel *C*; this motion is recorded by the

Fig. 3.

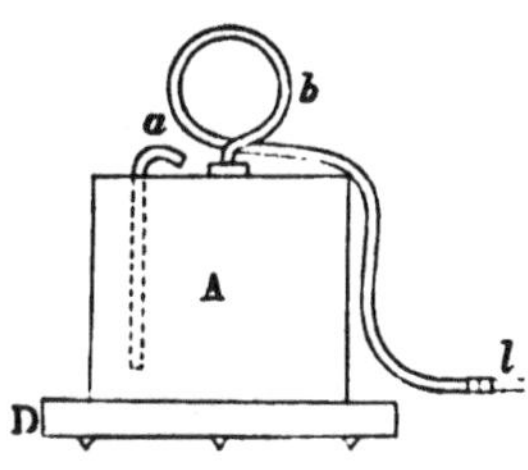

Fig. 4.

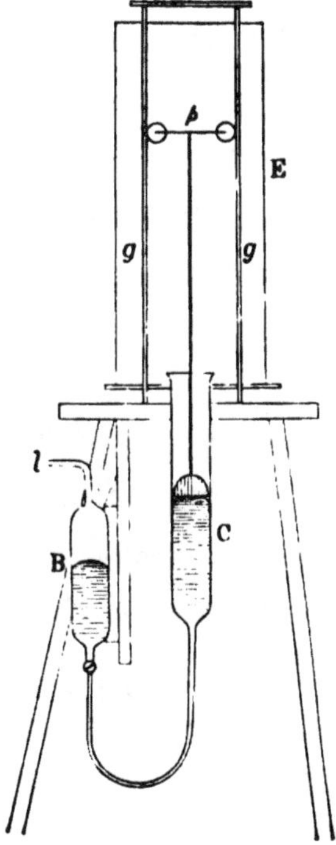

pen on the vertical cylinder revolving uniformly. The whole apparatus, which is constructed in a form specially adapted for transporting and setting, is given in the photograph No. 3 and 4 of the frontispiece.

The relation between the change of water level above the jar and the mercury meniscus in the tube can be found in the following way :—

Fig. 5.

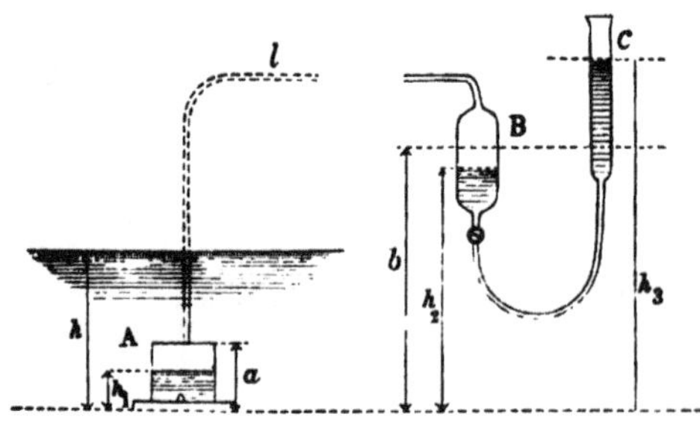

Let h, h_1 (Fig. 5.) be the levels of sea water above and inside the jar respectively; h_2, h_3 be the levels of mercury in the tubes B and C. Let b be the common height of the mercury in the two tubes, when the jar is not immersed in water. If π and P be the pressure of atmosphere and the pressure within the jar respectively, we have

$$P = \pi + h - h_1 = \rho(h_3 - h_2) + \pi ,$$

where ρ is the the density of mercury.

If s_1, a be the cross section and the height of the jar, and s_2, s_3 the cross section of the tubes B and C respectively, we have, by assuming Boyle's law,

$$P\{s_1(a - h_1) + v + s_2(b - h_2)\} = \text{const.},$$

where v is the volume of the copper tube plus that of the part of the tube B lying above the level b.

The differentials of the above two equations are

$$dP = dh - dh_1 = \rho(dh_3 - dh_2)$$
$$dP\,\{s_1(a-h_1)+v+s_2(b-h_2)\} - P(s_1dh_1+s_2dh_2) = 0\,;$$

we have also the equation of continuity

$$s_2dh_2 = -s_3dh_3$$

Eliminating dh_1, dh_2, dP from these equations, we get

$$\frac{dh_3}{dh} = \frac{s_1P}{\rho\left(1+\frac{s_3}{s_2}\right)\left\{s_1(a-h_1)+v+s_2(b-h_2)+s_1P\right\}+Ps_3}$$

Since the first three terms in the denominator of the above expression are very small compared with the fourth, we have, neglecting these small terms,

$$\frac{dh_3}{dh} = \frac{1}{\rho\left(1+\frac{s_3}{s_2}\right)+\frac{s_3}{s_1}}\,.$$

Hence the motion of the mercury meniscus in the tube C is practically proportional to the change of sea level. We may also infer from the exact expression of dh_3/dh that the volume of the copper tubing need not be small compared with that of the jar. Even, if the volume of the tubing is equal to that of the jar and the change of sea level exceeds 3 m., the actual value of dh_3/dh in the present apparatus does not differ from the value given in the last equation by more than 0.2%.

The effect of temperature may be calculated in a similar way. For this purpose, the product of the volume and the pressure in Boyle's law is put equal to RT, where T is the absolute temperature, and R a constant. Here h is considered to be constant; we have then the equations

$$dP = -dh_1 = \rho dh_3\left(1+\frac{s_3}{s_2}\right),$$

$$\rho dh_3\left(1+\frac{s_3}{s_2}\right)\left\{s_1(a-h_1)+v+s_2(b-h_2)\right\}-P(s_1dh_1+s_2dh_2) = RdT,$$

$$s_2\, dh_2 = -s_3dh_3 .$$

Eliminating dh_1, dh_2, dP, we have approximately

$$\frac{dh_3}{dT}=\frac{R}{P\left\{\rho\left(1+\frac{s_3}{s_2}\right)s_1+s_3\right\}}\doteqdot\frac{R}{s_1\rho P\left(1+\frac{s_3}{s_2}\right)}.$$

If we neglect v in comparison with the volume of the jar, R is equal to Ps_1a/T; hence

$$\frac{dh_3}{dT}=\frac{a}{\rho T\left(1+\frac{s_3}{s_2}\right)}.$$

In our case, $a=12$ cm., $\rho=13.6$ and $1+\frac{s_3}{s_2}=1.26$; hence at $10°C.$,

$$\frac{dh_1}{dT} = 0.0025 \text{ cm.}$$

When the temperature changes, the vapour tension also changes; but the change of vapour tension per degree rise of temperature in the range of ordinary temperature is about 1/3 the change of pressure due to the thermal expansion of air. Hence as the combined effect of these two, we may take

$$\frac{dh_3}{dT} = 0.0033 \text{ cm.}$$

We have also experimentally determined this ratio by heating the water in which the jar was immersed, and found 0.004 cm., which agrees fairly well with the above value. The greater part of the enclosed air, which is in the diving jar, is subject to the daily change of temperature by a few degrees of the sea bottom. It is also an easy matter to protect the remaining part of the air, from a considerable change of temperature.

Hence the error arising from the change of temperature is usually to be neglected.

To estimate the effect of the barometric change, both h and T are to be considered as constant; we have then

$$dP = d\pi - dh_1 = d\pi + \rho dh_3\left(1+\frac{s_3}{s_2}\right),$$

$$dP\{s_1(a-h_1)+v+s_2(b-h_2)\} - P(s_1dh_1+s_2dh_2) = 0$$

$$s_2dh_2 = -s_3dh_3.$$

Eliminating dh_1, dh_2, dP and neglecting small quantities, we get

$$\frac{dh_3}{d\pi} = -\frac{a-h_1}{P\rho\left(1+\frac{s_3}{s_2}\right)}$$

which is, in our case, nearly equal to -0.04 mm. for the change of pressure by 1 cm. of mercury. Hence the barometric change of 10 cm. in mercury only causes the displacement of the pen not amounting to ½ mm. Thus in actual case, the error due to the barometric change is quite insensible.

Thus the present tide-gauge may safely be used without incurring any sensible error due to the changes of temperature and pressure.

§ 3. GENERAL RESULTS.

In the excursion of the summer vacation of 1903, we were assisted by Messrs Y. Inouye, N. Watanabe and T. Hirata, then students of physics and now graduates. Stations were Niiyama and Ayukawa in Rikuzen; Inuboye in Shimôsa; O-maezaki and Maisaka in Tôtômi; Shionomisaki in Kii; Tei, Susaki and Yotsu in Tosa. Of these stations, Ayukawa and Susaki are bays, and others are open coasts.

In the excursion of the summer vacation of 1904, Mr. S. Iwamoto, a graduate of physics, was our cooperator; stations were the bays of Ryôishi and Kamaishi in Rikuchiu, the bays of Kojirohama, Yoshihama, Okirai, Ryôri and Ôfunato in Rikuzen, coasts of Niigata, Kashiwazaki, Naoetsu, Fushiki and Tsuruga on the Japan Sea coasts. Besides, Shizuura in Suruga, Kamezaki and Kamagôri in Mikawa were chosen as our stations. Susaki in Tosa was again observed in this summer.

In the winter vacation of the year, the excursion was undertaken on the coast of Kiushiu. Observed bays were Hososhima and Aburatsu in Hiuga, Kagoshima, Nagasaki and Shimonoseki. The bay of Hiroshima in the inland sea of Seto was also observed.

In the spring of 1905, observations were made at Atami on the western side of the Bay of Sagami. The summer vacation of the year was spent in the excursion to Hokkaidô; the observed bays were Otaru, Nemuro, Hanasaki, Hamanaka, Mororan and Hakodate. At Hakodate, Mr. Watanabe again assisted our observations. In the winter vacation, the bay of Shimoda in Izu was observed.

In the spring vacation of 1906, the bay of Moroiso in Misaki was observed. The summer vacation of the year was spent in the observation of the Strait of Naruto, which is famous for its rapid current; Mr. T. Fukuda was our cooperator. Since July, 1906 the tides of the Bay of Tôkyô have been recorded by a tide-gauge of our system, set up by the Office of the city.

In discussing our results, the valuable observations by Assistant Professor A. Imamura at Miyako and Ôdsuchi in Rikuchiu, and Hiragata in Hitachi were utilized. These observations were made by his portable tide-gauge; the duration of observation was generally a day and night.

In addition, the records obtained by Lord Kelvin's tide-gauges set up at the following ten stations were made use of: Hanasaki in Nemuro, Ayukawa in Rikuzen, Kushimoto in Kii, Hososhima in Hiuga, Fukahori in Hizen, Tonoura in Iwami, Wajima in Noto, Iwasaki in Mutsu, Kelung and Takow in Formosa.

The topographs of these stations and the records are given in the plates annexed to the present paper.

From the thorough study of the numerous records obtained, we may infer the following general conclusions:—

1. On the Pacific coasts free from any inlet in the coast line, the secondary undulation is very inconspicuous, and of a quite irregular nature.

2. On the Japan Sea side, the secondary undulation on open coast is conspicuous, though not regular.

3. In a bay of considerable area, or a shallow bay communicating with the ocean by a narrow opening, the secondary undulation is in ordinary cases inconspicuous.

4. In a deep bay or estuary, the breadth of which is not large in comparison with its length, the secondary undulations are most pronounced.

5. In bays or on open coasts, which are not far from each other, common undulation is often observed.

6. The secondary undulations in many bays often change their periods continuously, through certain ranges.

7. In some bays, the periods of the undulation are fairly constant.

8. In many cases, the same trains of secondary undula-

tions appear in the same phase with respect to the tidal wave on consecutive days of ordinary weather.

9. The phases of the prominent fundamental undulation at different parts of a bay are equal.

10. The periods T of the most pronounced undulations is fairly given by the relation

$$T = \frac{4l}{\sqrt{gh}}$$

where l is the length of the bay measured along its depth, h the mean depth of the bay, and g the acceleration due to gravity.

11. Just outside a bay, the undulation, which appears inside the bay with considerable amplitude, may also be traced, but its amplitude is very small.

12. In a bay, the periods of the conspicuous undulations in the case of a storm, or a sea wave of distant origin, are the same as those ordinarily observable in the bay.

It has long been believed that the secondary undulation in a bay is the seiches between two opposite sides of the bay, but according to our observations, the phase of the conspicuous undulation is usually the same throughout the bay, so that this view can not be generally true.

Napier Denison considers the undulation to be long waves propagating from the ocean toward the bay. All results above enumerated, except the 7th, 9th, 10th and 12th, can be explained by his view. But the fact, that there is a prominent undulation peculiar to each bay, can not be explained by merely considering progressive waves.

The undulation in a bay can, however, be explained in the following way. For the sake of simplicity, take a rectangular bay of a constant depth. Suppose a regular series of long waves propagated in the direction of the length of the bay and reflected at its end. By the interference of the incident and reflected waves, a stationary wave is formed, having its loop at the end of the bay. If the wave length be such as to form the node at the mouth of the bay, the period is the same as that of the fundamental oscillation of a tank having double the length of the bay, and therefore the amplitude of oscillation must necessarily be magnified by the successive incidence of the long wave. The case is just analogous to the acoustical resonance of the air column to the vibration of a tuning fork placed over its mouth. The period of the oscillation is then, neglecting the mouth correction, expressed by the relation

$$T = \frac{4l}{\sqrt{gh}}$$

where $\sqrt{gh}$ is velocity of the long wave.

If the waves of different periods proceed from the ocean toward the shore, the one whose period coincides with that of the oscillation having its node at the mouth of the bay, will excite the most energetic oscillation of the bay water. Thus bays on the coast line may be compared with a series of resonators, each of which takes up selectively and resonates to the note of its proper period from the chaos of very complicated sounds or noises from the exterior. The plausibility of such an idea seems to be established in a rather unexpected degree by the present investigation. Moreover, the fact that the motion of the level of a bay in the principal undulation is in the same phase for several stations, stands in favour of the above view. G. H. Darwin and Otto Krümmel* seem to have entertained an analogous idea.

*) Darwin, The Tides, Ch. X, p. 169. O. Krümmel, Ueber Gezeitenwellen, Rede bei Antritt d. Rektorates d. Königl. Christ.-Albr.-Univ. z. Kiel, 5 März, 1897.

In a bay, beside the uninodal oscillation above referred to, the oscillations with two nodes, three nodes etc., are equally possible; the periods of these oscillations are respectively $\frac{1}{3}$, $\frac{1}{5}$ etc. of the period of the fundamental oscillation. In some cases, the lateral oscillation of the bay excited by incident waves is also possible, the period of which is principally determined by the oscillating water in the bay. These modes of oscillations were actually found to exist in some bays such as Hososhima, Ôfunato, and Hakodate.

In the oscillation of the bay water just referred to, the period of the forcing wave which corresponds to the maximum resonance is not sharply defined; but within a certain range of the period, the oscillation remains fairly conspicuous, as we have often observed.

In bays of regular shape, such as Ôfunato and Hososhima, the position of the nodal line is determinate; but in bays of complicated shape, such as Shimoda and Susaki, several nodal lines are conceivable. By the choice of the nodal lines, the length and mean depth of the bay vary within a considerable range, so that the period of the proper oscillation changes within a certain range. Hence such a bay may resonate to any one of the incident waves with the period lying within the same range. In the two bays above mentioned, the period of the conspicuous undulation was actually found to vary within a moderately wide range.

As to the cause of the long waves, which manifest themselves as secondary undulations, we may mention the wind, the cyclone, the earthquake, etc. It is a matter of fact that the seiches in many lakes, which are the result of interference of a direct and a reflected wave of long wave-length, are often excited by wind. In the same way, the wind blowing on the surface of the ocean, may cause waves several kilometers long. Long waves of a considerable amplitude are often caused by a deep cyclonic centre. Near a cyclonic center, the fluctuation of pressure or of wind velocity are incessantly going on, and this acting in an impulsive way, may cause waves of long periods. An upheaval or depression of sea bottom due to an earthquake or to a submarine eruption, may also be a cause of sea waves of considerable periods.

§ 4. SPECIAL RESULTS.

In the present section, the results regarding each special bay will be described. Here it will be remarked that the number of observed periods are naturally rich at stations in which the observation was continued for a long interval, while they are poor at stations temporarily observed during one or two days. The general results, which we have summarized in the foregoing section, were deduced from the following results of observations.

I. COASTS OF HOKKAIDÔ.

(1) Otaru (Aug. 12–15, 1905). Top. 1. Pl. I, Fig. 1.

Otaru is a city situated on the northern coast of Hokkaidô. The observation was made on the middle shore of the harbour. Here, as in other coasts of Japan Sea, the tidal range is very small, not exceeding 30 or 40 cm.; the secondary undulation is conspicuous, but not very regular. On the western side of the harbour, Kelvin's tide-gauge is constantly working; comparing the record of the instrument with that of ours, we may conclude that the principal motion of the water in the harbour is in the same phase.

The observed periods are **$13.8^m-16.5^m$**, 24.3^m, 36.1^m, 45.0^m; in the above and in what follows, the heavily printed figures denote the conspicuous undulation. The calculated period 17.3^m fairy coincides with the observed period of the conspicuous undulation.

It should be remarked that in our observation, short waves of periods not exceeding a few minutos were always eliminated by properly adjusting the tide-gauge.

(2) Nemuro (July 28–31, 1905). Top. 2. Pl. I, Fig. 2.

Nemuro is a small bay on the eastern extremity of Hokkaidô; here the tidal range is tolerably large. The periods observed are **10.9^m**, 33.7^m, and 38.6^m; while the calculated period is 9.0^m.

(3) Hanasaki (July 31–Aug. 1, 1905). Top. 3. Pl. I, Fig. 3.

Hanasaki is a shallow inlet on the Pacific coast not far from Nemuro; here also Kelvin's tide-gauge is constantly working. The undulation is generally inconspicuous, but regular undulation of long period may often be traced: The periods observed are, 6.9^m, 14.2^m–17.7^m, 19.2^m–23.2^m, 38.8^m, 44.5^m and 61.2^m–65.6^m; the calculated period is 10.9^m.

(4) Hamanaka (Aug. 4–5, 1905). Top. 5. Pl. II, Fig. 1.

Hamanaka is a bay situated on the south-eastern coast of Hokkaidô. The observation was made at Kiritapp on the western side of the bay. During the observation, a low atmospheric pressure was approaching from the Pacific side of Honshiu to Hokkaidô, and a conspicuous, but not regular undulation of long period appeared on the record. Periods observed are 20.9^m, 24.2^m, **49.5^m**, 62.3^m and 84.3^m, while the calculated is 48.2^m.

(5) Mororan (Aug. 17–22, 1905) Top. 7. Pl. II, Fig. 2–3.

Mororan is a small bay on the southern side of the Volcano

Bay, situated near its mouth. The secondary undulation is conspicuous and fairly regular; observed periods are 46.7^m, **51.1–54.0^m**, and 58.3^m–60.0^m; while the calculated is 48.9^m.

(6) Hakodate (July 21–Aug. 10). Top. 8. Pl. II, Fig. 4–5; Pl. III.

Hakodate is the best anchorage in Hokkaidô, situated on the middle coast of the Strait of Tsugaru, which separates Hokkaidô from Honshiu. The bay is approximately of a semi-circular shape; here Kelvin's tide-gauge has been continuously working during the last 20 years, and has recorded several important waves accompanying the sea waves originated along the coasts of Pacific. Our observations were chiefly made at the innermost parts of the bay; i.e. near the wharf of Hakodate. The simultaneous observations between Hakodate and Kamiiso on the opposite sides of the bay, or Hakodate and Tachimachizaki, just inside and outside the bay respectively, were also carried out.

At Hakodate, the secondary undulation is very conspicuous, sometimes exceeding 30 cm. and fairly regular. The periods of the most conspicuous undulation range from **45.5^m** to **57.5^m**. Sometimes its octave **21.9^m–24.5^m** is found superposed on the undulation of the above period. In the undulation accompanying the sea waves, the octave always appears in a very marked degree. The longer period was found to correspond to the fundamental oscillation of the bay, while its octave corresponds to its lateral oscillation.

The calculated periods for these oscillations give 45.3^m and 23.6^m respectively. As we shall see hereafter, the periods of oscillations corresponding to these modes, as given by our model of the bay, are 47.0^m and 23.6^m respectively.

When the simultaneous observations between Hakodate and

Kamiiso were carried out, the fundamental oscillation only appeared on our records. The comparison of our records shows that the phase of the oscillation is the same for these two stations.

The simultaneous observations between Hakodate and Tachimachizaki showed that notwithstanding the conspicuous undulation appearing in the bay, the undulation just outside the bay was almost insignificant. This is a good example for illustrating our view regarding the secondary undulation, which we have propounded in the foregoing section. Beside the periods above described, a longer period of about 120^m is sometimes traceable.

The amplitude of secondary undulation is usually increased by a low barometric pressure approaching the bay. As a good example, we may cite a cyclone on Sept. 21–22, 1904, which was approaching from the Pacific side of Honshiu toward Hakodate. The undulation in the bay (Pl. III, Fig. 1) continued over a whole day with a considerable amplitude, its maximum exceeding 40 cm.; the periods of conspicuous undulation were 47.1^m–56.9^m and its octave.

The bay is especially sensitive to incident sea waves; waves originating on the American coasts were often beautifully recorded by the tide-gauge in the bay. The periods of the Ecuador wave in the bay were 21.9^m and 40.9^m–49.2^m; while those of the Valparaiso wave were 22.1^m and 48.0^m–53.0^m.

The periods of the great sea waves of Sanriku (Pl. III, Fig. 2) in 1896 were 18.8^m, 39.5^m, and 57.5^m; those of the small sea waves (Pl. III, Fig. 3) in 1897 were 22.1^m and 45.5^m.

II. JAPAN SEA COASTS OF HONSHIU.

(1) Aomori (July 18–21, 1905). Top. 9. Pl. IV, Fig. 1–2.

The large Bay of Mutsu has the form of a dumb-bell, connected to the Strait of Tsugaru by a wide neck. The observation was made at Aomori, when the centre of a deep low pressure was approaching the bay from Japan Sea side (Pl. XCIII). On this occasion, a regular undulation of 103^m appeared on the record and continued for a day and a half. Upon this, an undulation of the period ranging from 23.4^m to 26.3^m was superposed. Besides, a period of about 300^m may also be traced.

The undulation of 103^m is probably the lateral oscillation of the bay; the period 108^m calculated on this view is in fair agreement with the observed period. The undulation of the shortest period may be its higher harmonics. The longest period is perhaps that of the fundamental oscillation of the bay having its node at the mouth; the calculation gives 284^m as the period of this mode of oscillation.

(2) Iwasaki. Pl. IV, Fig. 3–4.

Iwasaki is situated on a northern coast of Mutsu; here Kelvins' tide-gauge is constantly working. The undulation in ordinary weather is generally conspicuous; in stormy weather (Pl. IV, Fig. 3–4), undulations of considerable amplitude are sometimes observed. The periods observed are 8.3^m, 11.0^m–13.5^m, $\mathbf{15.8^m}$–$\mathbf{17.3^m}$.

(3) Niigata, Kashiwazaki, and Naoetsu (Aug. 9–17, 1904). Pl. V, Fig. 1–2.

These stations are situated along the northern coast of Echigo. The secondary undulation in these coasts are conspicuous, but much complicated by the superposition of several minor components. The periods observed are: 22.6^m at Niigata; 11.6^m–12.4^m, 14.5^m–17.2^m, 22.0^m–25.5^m, 30.0^m–35.8^m and 43.3^m at Kashiwazaki; and 37.6^m at Naoetsu.

(4) Fushiki (Aug. 14–24, 1904). Top. 12. Pl. V, Fig. 3–4.

Fushiki is situated at the end of the Bay of Toyama; since 1905, a tide-gauge of our system has been set up and the record continuously taken, under the care of the meteorological station of the city. Here the secondary undulation is not conspicuous; but the regular undulation of **11.3^m–14.9^m** is sometimes observed. In addition to this, the periods 30.0^m, 54.1^m, 58,0^m and 119^m are to be traced. The period calculated is 52.7^m.

Here we had several remarkable examples of the instances, in which a strong wind blowing toward the land often heaps up the water on the shore to a considerable height.

(5) Wajima. Pl. V, Fig. 5.

Wajima is situated on the northern coast of Noto; there is a Kelvin's tide-gauge at work. In ordinary weather, undulation of considerable amplitude often appeared on the record.

The periods observed are 12.5^m–16.4^m, 21.9^m, 28.0^m–33.0^m, 81.5^m.

(6) Kanaiwa (Aug. 25–26, 1904). Pl. VI, Fig. 1.

Kanaiwa is situated on the open coast near Kanazawa; the secondary undulation is not only inconspicuous, but also irregular. The periods 21.2^m and 59.0^m may however be traced.

(7) Tsuruga (Aug. 28–30, 1904). Top. 13. Pl. VI, Fig. 2–3.

Tsuruga is a city at the end of a long V-shaped bay, which forms a part of the Bay of Wakasa. Usually the undulation is conspicuous, but not regular; undulations of the periods **56.7^m** and **62.7^m–67.7^m** often appear with considerable amplitudes. We can also trace shorter and regular undulations of the periods **10.5^m** and **22.6^m**.

The calculated period 59.0^m fairly agrees with the observed. It is however to be noticed that at the mouth of the bay, several nodal lines are conceivable, which may change continuously from one to the other. The above value corresponds to the mean position of the nodal lines, and therefore the calculated values corresponding to the other nodal lines, may be greater or less by several minutes than the above value. In the actual case, we also found the periods varying continously from 56.7^m to 62.9^m or from 62.9^m to 67.7^m.

(8) Tonoura. Top. 15. Pl. VI, Fig. 4–5.

Tonoura is a small inlet situated north of Hamada in Iwami; here a Kelvin's tide-gauge is at work. The secondary undulation is conspicuous, especially in stormy weathers (Pl. VI, Fig. 4–5). The characteristic period in the bay is **11.9^m–12.9^m** while the calculated value is 11.1^m, in good accordance with the actual. Besides, the periods 15.3^m and 21.5^m–28.8^m are sometimes observed.

In concluding the description regarding Japan Sea coast, it may be observed that in many of these stations, undulations of the periods 120^m–130^m and 150^m–180^m may sometimes be traced.

III. PACIFIC COASTS OF HONSHIU.

(1) Samé (Aug. 24–25, 1905). Pl. VII, Fig. 1.

Samé is an open coast on the Pacific side of Mutsu; here the secondary undulation is conspicuous, but not regular.

The periods observed are 16.4^m, 35.0^m and 41.5^m.

(2) Miyako. Top. 16. Pl. VII, Fig. 2–3.

Miyako is a bay on the coast of Rikuchiu, where a meteorological station is placed. Dr. A. Imamura has set up a tide-gauge of the Richards' type at Kajigasaki, for the purpose of investigating sea waves in connection with the earthquake, and many valuable records were obtained, some of which are reproduced in

Pl. VII. Fig. 2–3. The periods usually observable are 12.0^m, **21.3^m–22.0^m**, 23.0^m–27.6^m and 55.2^m. The periods observed in the bay in the case of a storm or a sea wave of distant origin are those usually found in the bay. The calculated period 24.0^m, which corresponds to the seiches between Miyako and the end of the bay, fairly accords with the conspicuous observed periods 21.3^m–22.0^m.

(3) Ôdsuchi (Aug. 11–12, 1902). Top. 17.

Ôdsuchi is a bay not far from Miyako. Here the undulation is conspicuous; the period observed by Dr. Imamura is 27.0^m, while the calculated period is $30.7.^m$

(4) Ryôishi and Kamaishi (July 21–28, 1904). Top. 18–19. Pl. VII, Fig. 4; Pl. VIII, Fig. 1–2.

The bays of Ryôishi and Kamaishi in Rikuchiu have their mouth in common and in reality form a W-shaped bay.

The principal station was chosen at Kamaishi at the end of the latter bay; other stations were Heida, Kamagasaki, Washinosu and Aodashi. The simultaneous observations in these stations taken three at a time showed that the phase of the principal secondary undulation is the same for these stations. The amplitude of the undulation did not much diminish at Kamagasaki and Washinosu as compared with the amplitude at Kamaishi. At Aodashi near the mouth of the bay, the secondary undulation was very inconspicuous. The undulations at Ryôishi were much complicated by the superposition of shorter waves. We could sometimes find the undulations of the same period and phase as those at Kamaishi. The periods observed are 12.0^m–13.2^m, **20.3^m** and **22.8^m** at Ryôishi, and 8.6^m–9.4^m, **20.3^m** and **24.8^m–26.0^m** at Kamaishi; the calculated periods for these bays are 21.3^m or 20.0^m and 24.8^m or 22.3^m respectively.

Dr. Imamura also obtained the undulation of the same periods in 1902.

The bay of Ryôishi is of a V-shape, and consequently all waves proceeding toward the bay are found at Ryôishi situated at the end of the bay. On the other hand, the bay of Kamaishi is somewhat crooked so that at Kamaishi near its end, the sea is extremely calm giving rise to few short waves such as are always observable in free coasts. A wave, whose wave length is very large compared with an obstacle, goes round the obstacle; but a wave whose wave length is very small, is screened by it and the sea behind it is quite free from its influence.

At Kamaishi and Ryôishi, we met with a storm on July 27–28; the character of the secondary undulations was not altered, except that they were much superposed by zigzags of shorter periods, and that their amplitude was somewhat increased.

(4) Kojirohama (July 26–31, 1904). Top. 20. Pl. VIII, Fig. 3.

Kojirohama is a small bay, south of the bay of Kamaishi; stations were Kojirohama and Ôishi, which are facing to each other. The simultaneous observation showed that the phase of the most prominent undulation is the same for these two stations. The observed periods are 18.8^m–20.4^m and 24.6^m, while the calculated period is 26.0^m.

It is remarkable to observe that the prominent period in the bay is nearly the same as that of Kamaishi or Ryôishi. Dr. Imamura also observed undulation of the same period in the bay.

(5) Yoshihama (July 28–Aug. 6, 1904). Top. 21. Pl. IX, Fig. 1–3.

This bay lies in the south of the bay of Kojirohama and not far from it. Stations were Yoshihama, Konpaku, Senzai and Kokabe. The secondary undulation was generally irregular and inconspicuous; but on Aug. 2–3, there appeared an unusual undulation of regular type of the period 18.5^m–19.6^m. Probably this undulation was connected with the low pressure then prevailing over the Pacific to the south of Tosa. It was also found that the undulation was very faint at Senzai and Kokabe near the mouth, and conspicuous at Yoshihama and Konpaku at and near the end of the bay respectively, and that the phases of the principal undulation were the same for these stations. The periods observed are 15.4^m, 16.5^m-17.9^m, **18.5^m–20.1^m**, 22.2^m-23.1^m, and 32.0^m-37.2^m. The calculated period is 21.1^m in good coincidence with the period of the conspicuous undulation.

(6) Okirai (Aug. 4–7, 1904). Top. 22.

Stations were chosen at Okirai and Koishihama. The bay of Okirai had a rather narrow mouth. The observed periods are 10.0^m, **27.5^m–29.9^m**, and 54.5^m, while the calculated period is 26.4^m.

(7) Ryôri (Aug. 8–9, 1904). Top. 23.

The bay of Ryôri has a form similar to the bay of Yoshihama, but its dimension is much smaller; the station was chosen at Nonomae.

The undulation was not conspicuous; but the periods 12.9^m, 18.3^m, 29.0^m and 33.3^m may be traced. The undulations observed by us were not conspicuous, though one of the periods lies fairly near the calculated period of free oscillation 18.4^m.

(8) Ôfunato (Aug. 8–10, 1904). Top. 24. Pl. X, Fig. 1–4.

The bay of Ôfunato has an elongated form and is somewhat crooked near its mouth, so that at Ôfunato situated at the end of the bay, the sea is extremely calm. It has a form specially fitted for the comparison of the phases of the secondary undulations at different stations along the bay. Stations were Ôfunato, Sunagosaki, Takonoura and Hosoura. The simultaneous observations between Ôfunato and other stations were taken at three different dates.

Fig. 1 and 3, or 2 and 4 are the records of the simultaneous observations; 1 and 1', 2 and 2', and 3 and 3' in these curves indicate the positions of the corresponding time. They clearly show that the phase of the conspicuous undulation is the same for these three stations.

The periods observed are 5,5^m, **12.8^m–16.8^m**, **36.0^m–39.1^m** and 41.5^m–43.5^m, while the calculated period is 36.4^m in good agreement with the observed. At Ôfunato and Hosoura where undulation of periods 12.8^m–16.8^m was sometimes observed, the phases of the undulation are opposite to each other. The undulation was very inconspicuous at Sunagosaki situated midway between Ôfunato and Hosoura so that it may probably be a binodal oscillation of the bay.

Waves of the short period 5.5^m appeared at Hosoura and Sunagosaki but not at Takonoura and Ôfunato; the absence of the waves at the latter stations is possibly due to the effect of shadow.

(9) Niiyama (July 21–Aug. 3, 1903). Top. 25. Pl. XI, Fig. 1–2.

Niiyama is small V–shaped bay; the undulation is not very conspicuous. The observed periods are **6.4^m–7.6^m**, 11.0^m–12.8^m, 20.0^m–23.7^m, 27.5^m–30.7^m, 61.5^m, 71.6^m and 90.0^m; while the calculated period is 7.5^m.

(10) Ayukawa (July 18–21, 1903). Top. 26. Pl. XI, Fig. 3; Pl. XII, Fig. 1–2.

Here a continuous record by Kelvin's tide-gauge has been taken since some ten years ago; many valuable records connected with sea waves and low pressures were obtained. Our observation was made at the tide-gauge station; the observed periods are **6.8^m–8.9^m**, 14.2^m–15.1^m and **20.9^m–22.8^m**; while the calculation gave 8.9^m. The same periods were also found in the case of several sea waves accompanied by earthquakes or atmospheric low pressure.

(11) Shiogama and Hiragata.

From the records obtained by Dr. Imamura at Shiogama in Rikuzen, and Hiragata in Hitachi, we found the following periods:—

44^m at Shiogama,

28^m and 50^m at Hiragata.

(12) Cape of Inuboye (Aug. 6–17, 1903). Pl. XII, Fig. 3.

The observation was made at the Cape of Inuboye. The secondary undulation was not conspicuous, being superposed by the short waves of considerable amplitude; the observed undulations were **8.9^m**, 16.3^m–18.1^m, 20.0^m, 26.0^m–29.8^m, 31.0^m–34.4^m, 38.7^m, 49.0^m and 66.0^m.

(13) Tôkyô. Top. 29. Pl. XIII, Fig. 1–4.

On the coast of the Bay of Tôkyô, several tide-gauges of Kelvin's type are constantly working. Our tide-gauge was set up at Etchiujima and records were continuously taken from Nov. 24 to Dec. 5, 1904. Since August, 1906, a tide-gauge of our system has been placed at Kanegafuchi on the bank of the River Sumida about 2 km. distant from its mouth. In usual weathers, the undulation of the bay is very faint, but regular and characteristic undulations of **63.1^m- 67.0^m** and 72.0^m–82.1^m are sometimes observed. In addition, the long periods

110^m–130^m are often traceable. Since the bay is very large and shallow, it can not easily be set in oscillation as a whole by any usual cause of excitement. Judging from the calculated periods 222^m and 158^m for the fundamental and the seiches oscillation respectively, the observed periods may be the higher harmonics of these oscillations, but considering the smallness of the amplitude of oscillation, they are perhaps rather due to progressive waves.

In the record at Kanegafuchi, the relation between the rise of the level by a flood and the change of the tidal range deserves notice. As the level of the river increases, the tidal range (Pl. XIII, Fig. 4) becomes gradually less and at last very small. As the level gradually falls the tidal undulation is again restored.

(14) Moroiso. Top. 30. Pl. XIV, Fig. 1–5; Pl. XV, Fig. 1–4.

About 4 km. north of Misaki, there lies a small branched bay; the one branch is called Moroiso and the other Aburatsubo.

At Aburatsubo, a Kelvin's tide-gauge is constantly working, to the record of which Professor F. Omori has frequently referred as Misaki mareogram. In the spring vacation of 1906, we also made simultaneous observations at different parts of the bay.

The undulation is very regular and conspicuous, having the periods **13.8^m-15.6^m**; the calculation gives a fairly coincident value 13.4^m.

The record shows an appearance of the beat of two waves of nearly the same wave length. So it was suspected that the phenomenon may be due to the interference of the two distinct modes of oscillation of the two branches of the bay which constitute a vibrating system with two degrees of freedom; but this is not the case, since the simultaneous observations at different parts of the bay showed the identity of waves with respect to their forms and phases. Pl. XIV, Fig. 2 and Pl. XV, Fig. 1, or Pl. XIV, Fig. 4 and Pl. XV, Fig. 2 are the records of the simultaneous observations at Moroiso and Aburatsubo. Pl. XIV, Fig. 5 and Pl. XV, Fig. 3 are the records of the simultaneous observations inside and outside the bay; placing the one record upon the other, we can distinctly trace undulations in the two records, which correspond to each other. The amplitude of the wave outside the bay is however very small, as compared with that of the undulation inside the bay. Pl. XV, Fig. 4 is a record at Aburatsubo in a stormy weather.

(15) Atami (April 2–7, 1905). Top. 31. Pl. XVI, Fig. 1–3.

Atami is a town situated on the western side of the Bay of Sagami and famous for the geyser. The secondary undulations were so inconspicuous as to make it difficult to detect their periods. We can, however, sometimes trace on the records the periods 12.8^m, 72.4^m–76.2^m and 97.6^m.

(16) Shimoda (Jan. 4–8, 1906). Top. 32. Pl. XVII, Fig. 1–3.

Shimoda is a small harbour at the southern end of Izu peninsula. The amplitude of the secondary undulation is so conspicuous that it is generally known as *yota*. The oscillation becomes conspicuous, when a centre of low pressure is approaching from the Pacific towards the place. The periods observed are 11.9^m, **13.8^m–18.2^m**, 21.5^m and 30.9^m.

In the bay, two extreme nodal lines are conceivable, the periods corresponding to these lines are calculated to be 13.3^m and 15.9^m. Hence any wave of the period lying between the values 13.3^m and 15.9^m may excite the oscillation of the bay; the periods actually observed fall nearly within the same limits.

(17) Shizuura (Sept. 4–6, 1904). Top. 36. Pl. XVIII, Fig. 1.

In Shizuura near the end of the Bay of Suruga, the secondary undulation is not conspicuous. The periods observed are **18.1^m–19.6^m** and 71.0^m. The calculated value is 54.5^m, so that the conspicuous undulation may be a binodal oscillation of the bay, but the amplitude of the wave being very small, the wave is possibly a progressive one.

(18) Omaezaki (July 31–Aug. 5, 1903). Pl. XVIII, Fig. 2.

The station is situated on a south-eastern corner of Tôtômi; the undulation is very inconspicuous. The periods observed are 18.6^m and 27.7^m.

(19) Maisaka (July 17–28, 1903). Pl. XVIII, Fig. 3.

Maisaka is situated on an open coast of the same province. The undulation is inconspicuous; the observed periods are 10.0^m, 16.0^m, 20.2^m–23.4^m, 30.2^m and 55.0^m.

(20) Kamagôri (Sept. 1–3, 1904). Top. 38. Pl. XIX, Fig. 1.

Kamagôri is situated near the eastern end of the Bay of Mikawa; this bay has a considerable area, and is very shallow, so that its oscillation can not be easily excited by any ordinary cause. The undulation is not very conspicuous, observed periods are **18.7^m**, 36.5^m and 43.2^m–45.1^m. An undulation of long period of 208^m may also be traced.

(21) Kamezaki (Aug. 31–Sept. 2, 1904). Top. 38. Pl. XIX, Fig. 2.

Kamezaki is situated near the northern end of the same bay. Inconspicuous undulations of 44.5^m and 68.0^m are observed; a period of 390^m may also be traced.

During our observation, we met with a storm, but the amplitudes of the slow undulation were not much affected by it.

The Bay of Mikawa and the Sea of Ise form a large connected system. Three modes of oscillation are possible: (*a*) the first and gravest mode is the oscillation of the system as a whole, (*b*) the second is that of seiches between the Bay and the Sea, and (*c*) the last the oscillation of the Bay only.

The calculated periods corresponding to these modes of oscillation are respectively 363^m, 278^m and 217^m. Though long periods obtained from observation fall near the calculated values, it is difficult to decide, from the scanty data, whether the observed periods actually correspond to the modes of oscillations above referred to.

(22) Shionomisaki (July 29–Aug. 7, 1903). Pl. XIX, Fig. 3; Pl. XX, Fig. 1–3.

Shionomisaki is the foremost promontory of Kii, where a light-house is placed. The observation was made at a beach below the light-house; the secondary undulation is very inconspicuous, but we can sometimes trace the undulation of the periods 11.3^m–16.3^m, 25.8^m and 34.1^m.

In a small bay of Kushimoto, which is not far from the promontory, Kelvin's tide-gauge is constantly working. It is remarkable to observe that while the secondary undulation at the promontory is inconspicuous, the undulation in the bay is very conspicuous. The periods **11.6^m-13.0^m, 16.5^m-$18\ 6^m$, $21\ 5^m$-23.7^m** and 32.1^m are observed with considerable amplitudes, especially in connection with the cyclones (Pl. XX, Fig. 2-3). The calculated periods 12.8^m and 18.3^m fairly coincide with the observed. The tide-gauge frequently recorded the sea waves originated on the American coast of the Pacific.

The periods observed in the case of a cyclone or of sea waves are the same as those observed in ordinary case.

(23) Bay of Ôsaka (Aug. 2-15, 1902). Top. 43. Pl. XXI-XXV.

Before the present investigation was commenced, the oscillation of the Bay was observed by the seiches-party of the Committee, S. Nakamura, K. Honda, Y. Yoshida, S. Iwamoto with Dr. Nakamura's limnimeters.

For the convenience of reference, the result is included in the present paper. The Bay has the form of an ellipse and communicates by two necks Akashiseto and Yuraseto to Harimanada and the Pacific respectively.

Stations were chosen at Imazu, Kishiwada, Yura, and Iwaya, and observations were made simultaneously. As shown in the topograph, these stations are evenly distributed along the shore line of the elliptical bay.

The original records, in which waves of short periods are beautifully traced (Pl. XXIII, XXIV, XXV), are not convenient for detecting the secondary undulations with long periods of several tens of minutes, or for studying the tide itself. For this purpose, the original records were reduced to a proper scale, some examples of which are given in Pl. XXI and XXII. Since the tide in Harimanada is nearly in opposite phase to that of the Pacific, the tidal wave at the stations is not naturally simple. Comparing the tidal phases at these four stations, it may be concluded that as we go from Yura inwards along the eastern coast of the bay, the tidal phase is gradually retarded, and at Kishiwada, the difference amounts to about 20^{m} to 75^{m}. The phase of the tide at Imazu is not however much retarded on the average as compared with that of Kishiwada. At Iwaya, facing Akashiseto, on the two sides of which the tidal phase is considerably different, the tide is naturally very complex.

Thus, as compared with the tide at Yura, the phase of the

tide at Iwaya is sometimes accelerated and sometimes retarded by several tens of minutes. The tidal wave in the bay chiefly enters through Yuraseto; but the Harimanada component, which enters through Akashiseto, is easily to be traced.

As regards the secondary undulations, the longest waves found in the records, have periods of 260^m-310^m; the amplitude exceeded ten cm. in one case, (Pl. XXII, Fig. 1-3). The phase of oscillations for the four stations, is the same, so that this undulation is probably due to the oscillation of the whole basin as a bay with its narrow necks at Akashiseto and Yuraseto. The result of calculation of the period for the mode of oscillation gives 270^m, which agrees well with the observed value. Along with the long wave, one with the period of ca. 100^m-140^m is often recognized; the phase of this undulation is generally the same for the northern stations Iwaya, Imazu and Kishiwada, while for Yura at the southern end of the bay, it is apparently opposite, in so far as we may judge from the faint traces of this component. It is possible that under favourable conditions, this undulation may become very prominent, though in the records at hand, it is scarcely to be detected. The undulation is probably due to the uninodal seiches of the bay along its longer axis. The calculated period of 120^m for the suspected seiches is in good agreement with the observed period. Besides, conspicuous undulations with the period of 50^m-65^m are often recognized. Comparing the phases of the undulation at different stations, it is at once found that for Kishiwada and Imazu, the motion is always opposite, and also that the phase of Yura is the same as that of Imazu. The undulation is probably that of the binodal seiches of the bay between Imazu and Yura. The calculated period of

60^m for this mode of oscillation falls fairly within the range of the observed values.

As to the secondary undulations of shorter periods, it may be generally observed that for each of the four stations, conspicuous waves with the periods 8^m–18^m are found very frequently. Periods of 20^m–25^m and 32^m–36^m are also met with. For Imazu and Kishiwada, waves of 8^m–18^m periods occur very conspicuously; and apparently idential trains of waves may often be traced at the two stations. Comparing the records of the four stations on the same days, waves common to different stations are often recognized. For examples :—

	Kishiwada	Imazu	Iwaya	Yura
Aug. 2	13.0^m	—	13.6^m	—
	9.4	—	9.8	—
3	11.2	—	13.6	12.7^m
8	22.0	22.8^m	—	—
9	12.7	12.1	—	—
	10.1	10.1	—	11.4
	20.6	21.6	—	—
	8.9	9.0	—	—
13	15.5	15.6	14.8	—
	13.3	13.6	—	—
14	16.2	16.3	16.6	—

Whether these waves are due to some higher modes of stationary oscillation of the sea, or they are merely progressive waves generated by meteorological or other causes is a question still to be solved.

Waves with periods shorter than about 8^m are also generally met with widely varying periods and often with considerable amplitudes.

For Kishiwada and Imazu, the short waves are generally very complicated, periods of 1.4^m– 1.5^m, and 2.0^m– 2.5^m being very frequent.

The records of Iwaya, when compared with those of the other stations are characterized by the simplicity of the short waves, the most prominent waves being of two groups, i.e. 1.0^m–1.3^m and 2.1^m–2.5^m.

(24) Bay of Hiroshima (Jan. 17–23, 1904). Top. 44. Pl. XXVI, Fig. 1–2 ; Pl. XXVII, Fig. 1.

The observation was simultaneously made in a small bay of Yedajima and in the harbour of Ujina.

In Yedajima, the tidal curve was often accompanied by a regular and inconspicuous undulation of 60.0^m. The period is far greater than the value calculated from the dimensions of the small bay of Yedajima. In Ujina, the undulation of the same period and phase as in Yedajima was also observed ; hence the undulation which was observed at Yedajima and Ujina, is probably due to the oscillation of the bay of Hiroshima as a whole. The calculation according to this consideration gives a fairly coincident value 61.6^m.

Beside this undulation, an oscillation of a shorter period 9.5^m was occasionally found both at Yedajima and Ujina.

The tidal wave in the bay undergoes a considerable change both in its form and phase, as compared with the tide on the Pacific coast. Here the tidal range is considerably large, and its form in the neighbourhood of its maximum or minimum is comparatively steep, as it is usually the case in a deep inlet. The retardation of the tidal phase is about 3 hours as compared with the tide on the Pacific coast of Shikoku.

(25) Shimonoseki (Jan. 14–17, 1905). Pl. XXVII, Fig. 2.

As it is to be expected, the tidal curve at Shimonoseki has a peculiar character, superposed by the secondary undulations of short duration of the periods 46.5^m, 54.3^m–57.6^m and 64.8^m. We can also trace an undulation of longer period of about 150^m.

IV. PACIFIC COASTS OF SHIKOKU.

(1) Strait of Naruto (Aug. 1–25, 1906). Top. 43. Pl. XXVIII–XXXII.

The Strait of Naruto is famous for its rapid current and the eddies accompanying it. The strait separates Harimanada from the Pacific by a narrow neck about 1.1 km. wide. The phase of the tide in the sea is just opposite to that of the Pacific; and when the Pacific is in the high water or in the low water, the sea is in the low water or in the high water respectively, so that at the strait, a level difference of 1 to 1.5 m. is produced, and consequently a torrent of water rushes from the ocean into the sea or in the reverse direction, according to the tidal phase. When the current attains its maximum velocity, it often exceeds 10 knots per hour. The current is always accompanied with roaring sound and eddies. Eddies are usually formed behind the stream; their diameter exceeds 6 meters, and they have funnel-shaped surfaces. If a boat be drawn into them, it is very difficult for it to get out.

The observations were made at 5 stations—Shioyasumi, Hinoura, Yebisujima Ôgeura and Kameura, of which different pairs were simultaneously observed. The records at Shioyasumi and Kameura give the tide on the Harimanada side, while those at Hinoura and Yebisujima give the tide on the Pacific side. The tide at Ôgeura is affected by these two.

The tidal curves of tho strait are much complicated by the influence of the two tides in opposite phases and by the superposition of the secondary undulations.

The records of Shioyasumi show a characteristic feature (Pl. XXVIII, Fig. 1–3); the tides at Hinoura (Pl. XXX, Fig. 1) and Yebisujima (Pl. XXIX Fig. 2, 4) are not simple.

At Ôgeura, which communicates with Harimanada and the Pacific, the tidal curve (Pl. XXX, Fig. 2, 3) has a very peculiar form. As regards the high or low water at this station, it agrees roughly with that on the Pacific side.

In these four stations, conspicuous, but not regular, undulations are observed, the periods of which are

63^m, 96^m–120^m	at Shioyasumi.
43^m, 54^m–59^m, 75^m–86^m, 111^m–121^m	at Hinoura.
51^m–63^m, 68^m–74^m, 84^m, 94^m–116^m	at Yebisujima.
54^m–57^m, 120^m, 180^m–200^m	at Ôgeura,

The tide of Kameura, which is about 1 km. distant from Shioyasumi and in the inside of the strait, is comparatively simple. The secondary undulations are very inconspicuous, though we can sometimes trace the undulations of the periods 16^m–20^m and 51^m–64^m.

Comparing the phases of the tide just inside and outside the strait, we observe that they are nearly opposite to each other (Pl. XXIX. Fig. 1, 2; Pl. XXVIII, Fig. 3; Pl. XXX, Fig. 1.) a fact which at first sight appears very curious. Now, Harimanada is connected to the other seas by three necks, namely Naruto Strait, Bisanseto and Akashiseto, of which the first, being very narrow as compared with others may be put out of consideration. The western neck Bisanseto is much wider than the eastern neck Akashiseto, the cross section of the former being more than twice that of the latter. Moreover the tidal range at Bisanseto is much greater than that at Akashiseto so that the flux of the sea water through the former neck will probably amount to more than three times that through the latter. Thus sea-level in Harimanada is principally determined by the tide from the western neck. The tidal wave of the Pacific enters the inland sea of Seto through the channel of Bungo, propagates eastward through the seas of Iyonada and Bingonada and arrives at Bisanseto, so that it requires about 5 hours to travel through the distance. This wave takes still 40 or 50 minutes to get at the strait of Naruto, so that inside and outside the strait, the phase differs by about 6 hours, as actually observed.

It is a matter of considerable interest to compare the phases of the components constituting the tidal wave on both sides of the strait. By means of a rectifier constructed by one of us,* we may eliminate the semi-diurnal or diurnal component, if it be known to exist. In this way, we may resolve a tidal wave into a certain number of components. Thus, the tides at Hinoura (Pl. XXX, Fig. 1) and Shioyasumi (Pl. XXVIII, Fig. 3) were respectively resolved into three components as shown in Pl. XXXI, Fig. 1, 2. They clearly show that the principal semi-diurnal components have opposite phases for Hinoura and Shioyasumi. In a similar way, the tides at Kameura (Pl. XXIX, Fig. 3) and Ōgeura (Pl. XXX, Fig. 3) were respectively

*) T. Terada, A Tide rectifier, Publications of the Earthquake Investigation Committee in Foreign Languages. No. 18, 1904.

resolved into three components shown in Pl. XXXII, Fig. 1, 2. The principal semidiurnal components at Kameura and Ôgeura have nearly opposite phases to each other. It is to be observed that in these figures, the diurnal components were not separated; they are very conspicuous at Kameura and Ôgeura, but not at Hinoura and Shioyasumi.

When the current was rushing from the Pacific into Harimanada an interesting phenomenon was observed. As the current increased its velocity, a regular undulation of about 2.5^{m} (Pl. XXVIII, Fig. 4) became gradually conspicuous and attained a maximum amplitude of about 18 cm., and then gradually decreased with the diminishing velocity of the current. In the records of Shioyasumi (Pl. XXVIII, Fig. 1–3), thickly zigzaged portions indicate the existence of such waves. Thus it seems very probable that the current behaves like a jet of air blown into the mouth of an organ pipe, causing standing oscillation of the air column in the pipe. The torrent of water rushing from the Pacific into the strait excites a standing oscillation of the water in the neighbourhood of the Strait. A few years ago, Professor H. Nagaoka expressed the possibility that the *Kuroshiwo*, which is the current along the coast of Japan with a velocity of a few knots per hour, may be the origin of the long waves observable on our coasts. The present case affords a good example for illustrating the above view on a small scale.

If we bring two records of consecutive days of normal weather into coincidence as regard the tidal phase, we observe in each record, the same succession of the secondary undulations following one after another; this remarkable fact is also noticeable on some other coasts.

(2) Tei (July 23—Aug. 3, 1903). Pl. XXXIII.

This station is situated on the Pacific coast of Tosa. The records show undulations of the periods 7.5^m, 25.9^m, 30.6^m–32.9^m, 39.9^m, 49.1^m–52.9^m and 73.9^m–77.5^m, small in amplitude and short in duration.

On July 30 to 31, the sea was very rough notwithstanding calm weather, when a remarkably simple secondary undulation of considerable amplitude appeared, lasting over 12 hours with a mean period of 51^m. The appearance of this remarkable undulation is probably connected with the low atmospheric pressure then prevailing over the vicinity of Formosa (Pl. XCIII).

(3) Susaki (Aug. 9–28, 1903 and Aug. 29–Sept. 7, 1904). Top 45. Pl. XXXIV–XXXVII; Pl. XXXVIII, Fig. 1–2.

Susaki is a deep bay on the middle coast of Tosa. The observations were made at 5 stations Yamasakibana, Shiraiwa, Ôtani, Heshima and Kure. The diagrams of Yamasakibana are characterized by simplicity of undulation, the periods observed are **30.9^m**, **35.4^m–38.5^m**, **39.9^m–41.6^m**, **43.1^m–46.8^m** and **50.0^m–54.0^m**.

As an illustration of the modes of superpositions of a series of different waves, rectified diagrams are given in Pl. XXXV, A. B. C. In curve A, a train of waves of the period 38^m is superposed on waves of the period 76^m, the former component being most conspicuous between *bc*, while the latter is pronounced between *ab*. In curve B, the waves of the period 76^m are superposed on the waves of the period 38^m between *ab*, the latter component gradually passing into the waves of 36^m between *bc*. In curve C, between *ab*, waves of the periods 51^m and 35^m form an apparent beat, and between *bc*, waves of 100^m are superposed on waves of 50^m.

When different rectified curves are compared with one another, the remarkable fact is revealed that almost identical forms of waves occur very frequently during the course of successive days. The examples of such a coincidence of wave-form are given in Pl. XXXVI, curve D–J. It appears that some particular form of waves is often repeated at the corresponding part of the tidal curve, for two consecutive days; this is shown in curve D and D′, in which identical waves occur in the same relation to the tidal phase. It also occurs that the same train of waves is recognized at the low water of one record and at high water of another, as shown in curve E and E′, or F and F′. When, however, different records, which are several days apart, are compared, the same waves are found apparently with no definite relation to the tidal phase as shown in curves H and H′. Curves I and I′ show an example in which a train of wave occurs on different days in apparently inverted form. In curve J′, the direction of the time is inverted. Curve K shows an example in which secondary waves of shorter periods are very faint, whereas a long wave of about 100^{m} is rather conspicuous. Curves L and L′ which are the rectified records of Yotsu, are given for comparison with K.

At Shiraiwa, the sea is very calm, whereas the secondary undulation of the period 31.0^{m} (Pl. XXXVIII, Fig. 1) is most pronounced. The periods observed at Ôtani are **17.6^{m}–18.2^{m}**, **35.4^{m}** and 53.3^{m}, and those at Heshima are **24.6^{m}–27.6^{m}**, **39.7^{m}** and 55.1^{m}. At Kure, secondary undulation of the periods **15.0^{m}–16.3^{m}** and **61.3^{m}** are noticeable.

Comparing the records of Kure and Yamasakibana on the same days (Pl. XXXVII, Fig. 3–4), it will be seen that while the wave somewhat longer than 60^{m} is common to both stations, the shorter waves of 35^m–42^m period appear only in the latter station, and those of 16^m only in the former. Again, comparing the records of Yamasakibana and Ôtani on the same days (Pl. XXXVII, Fig. 1,2), 35^m–42^m waves are found common to both stations, while 18^m waves are peculiar to the latter station. At Heshima, which is situated at the mouth of the minor inlet of Nomi, the 16^m wave is absent, and 35^m–42^m waves are apparently traced, though not of such great amplitude as at Yamasakibana.

Thus we may infer that the undulation of 35^m–42^m has a node near Kure, and that the period of about 16^m at Kure seems to be due to the oscillation of the minor inlet. The undulation of about 18^m is probably due to the seiches between Awa and Ôtani.

Calculated values of the periods corresponding to these supposed modes of undulation show a fair agreement with the actual periods.

As will be seen soon after, investigations with models also lead to the same conclusion.

(4) Yotsu (Aug. 28–30, 1903). Pl. XXXVIII, Fig. 3.

This station is situated on the western part of the Pacific coast of Tosa; the secondary undulations were very inconspicuous. Periods of 14.6^m–15.4^m and 75.8^m could however be traced.

V. COASTS OF KIUSHIU.

(1) Hososhima (Dec. 21–26, 1904). Top. 56. Pl. XXXIX–XL.

Hososhima is an elongated bay on the eastern coast of Hiuga; here Kelvin's tide-gauge is constantly working. It has recorded several sea waves, which were originated on the American coasts and traversed across the Pacific. Simultaneous observations were made at Hososhima and Isegahama situated inside and outside the bay respectively. In the bay, extremely regular undulations appeared superposed on the tidal waves; the periods varied from **17.8^m** to **20.3^m** according to the tidal phase. In calm weather, the amplitude of undulation amounted even to 25 cm. The period of the undulation slightly decreases, as the tide passes from low water to high. The calculated period is 19.0^m, which fairly agrees with the observed one. The change of the period caused by the change of the depth by tidal motion has also the range, which is to be expected from the theory. Besides, longer periods 34.0^m– 38.7^m and 43.4^m– 49.1^m are sometimes observed.

Outside the bay, undulations of the periods 17.8^m– 20.3^m are very faint, and inconspicuous undulations of the longer periods are also observable. Placing a record in the bay upon the corresponding one of the open coast, we can distinctly trace undulations in the two rocords, which correspond to each other.

On the open coast of Hososhima, we actually met with two small inlets some ten meters long, each of which was constantly excited by waves of short periods proceeding one after another toward the inlets, and made an approximate standing oscillation of considerable amplitude, having the node at its mouth.

If we bring two records of any consecutive days into coincidence as regards the tidal phase, we observe the same succession of undulations.

Pl. XL, Fig. 1–3 are the records in stormy weather obtained by Kelvin's tide-gauge. In the first two curves, the period corresponding to the binodal oscillation of the bay is very conspicuous while in the third curve, the periods of fundamental oscillation only are observable.

(2) Aburatsu (Dec. 27–31, 1904). Top. 58. Pl. XLI.

Aburatsu is a small bay on the southern coast of Hiuga; simultaneous observations were made at Aburatsu and Umegahama, inside and outside the bay respectively. In the bay, we observed conspicuous undulations, though they are not regular. The periods observed are **15.0^m–19.0^m**, 21.6^m–24.5^m, 37.5^m–39.2^m and 43.0^m; they are nearly the same as those observed at Hososhima. In ordinary weather, the amplitude of the conspicuous undulation often exceeds 13 cm. The calculated period is 15.1^m.

Outside the bay, the undulations are very inconspicuous; but the same periods as those inside are also traceable. If we bring two corresponding records in and outside the bay into coincidence, we can distinctly trace the corresponding undulations of one record in the other (Pl. XLI, Fig. 1 and 3, or 2 and 4). As in the case of Hososhima, if we bring two records of any consecutive days into coincidence as regards the tidal phase, we notice the same succession of secondary undulations.

(3) Kagoshima (Aug. 1–7, 1905). Top. 59. Pl. XLII–XLIII.

The large Bay of Kagoshima has an elongated form, 77 km. long and 20 km. broad in its widest part. The island of Sakurajima, an active volcano, situated about 11 km. from its end, divides the bay into two portions communicating with each other by two channels on both sides of the island. The observation stations were Kajiki, Kagoshima and Ibusuki; they are situated at the end, along one of the channels and at about 10 km. from the mouth of the bay respectively. The first and second stations, as well as the second and third stations were simultaneously observed.

At Kajiki, the tidal curve was extremely smooth, showing no trace of secondary undulation. It appears then that Sakurajima nearly screens long waves from being propagated into the bay. The tidal curves at Kajiki and Kagoshima almost coincided with each other, showing that the narrow channels have neither damping nor retarding effect for the oceanic tide of extremely long wave length.

The record of Kagoshima was generally very simple, but frequently regular undulations of 17.2^m and 22.8^m–23.9^m are observed.

At Ibusuki, the observed periods are 14.2^m and **18.3^m–20.6^m**. Since the undulations at Kagoshima were very faint during our observation, the existence of the corresponding waves at the station is not certain.

In the calculation of the period of the oscillation in the bay that part which extends from the mouth to Sakurajima, is only to be taken into consideration, because the rest is not disturbed by ordinary waves. The calculated period of the fundamental oscillation is 107^m, which exceeds very much the observed value. The observed undulation may possibly be due to progressive waves, which have the same periods as those frequently found in different coasts of Kiushiu. That the amplitude of the undulations is considerably less in Kagoshima than in Ibusuki, indicates the plausibility of the above view.

(4) Nagasaki (Jan. 9–13, 1905). Top. 60. Pl. XLIV–XLV.

Nagasaki is a well known harbour on the western coast of Kiushiu. The observation was made near the end of the bay. Since March 1905, a tide-gauge of our system has been set up in the same place by the Office, and many beautiful records obtained.

In the bay, the secondary undulation is so conspicuous that it is usually known as *abiki*. The observed periods are $22.5^m-25.2^m$, $\mathbf{31.9^m-32.4^m}$, $\mathbf{34.5^m-37.6^m}$, 40.1^m, $44.5^m-45.2^m$, 53.6^m and $69.0^m-72.0^m$; the amplitude of the conspicuous undulations often exceeds half a meter. On one occasion, about 10 years ago, the amplitude of the *abiki* was over 2 meters, and a large number of boats and steamers are said to have been damaged. The largest amplitude since the beginning of the tide-gauge observation was 1.54 m., at midnight on May 1, 1905. Pl. XLIV and XLV are records of the famous *abiki* together with records of less remarkable undulations.

The conspicuous *abiki* is generally associated with weather in which the isobars in the neighbourhood of the district is much crooked by two coexisting low barometric centres (Pl. XCIII—XCIV). It is well known that a tornado is frequently associated with such a distribution of isobars; then it seems very probable that a sudden local disturbance of pressure may often be accompanied with an unstable baromometric distribution. The *abiki* is often found in apparently calm weather; a deep barometric centre with regular concentric isobars, which is approaching the district, excites short waves of considerable amplitude, but does not cause the *abiki* in a marked degree.

As to mode of oscillation of the bay, two are conceivable. The one is seiches between Fukahori and the end of the bay; the other is the fundamental oscillation having the node at the mouth. The periods calculated on this supposition are 22.6^m and 37.5^m respectively, closely agreeing with the observed periods;

experiment with the model gave also fairly coincident values.

(5) Fukahori. Top. 60. Pl. XLVI.

Fukahori lies near the western mouth of the bay of Nagasaki; here Kelvin's tide-gauge is always working. The undulation is generally inconspicuous; the same periods as those observable at Nagasaki are also traceable. It is interesting to notice that though the ordinary oscillation of the bay is very prominent at Nagasaki, it is not conspicuous at Fukahori, the latter being situated at the node of the oscillation. Even the great *abiki* of Nagasaki on May 2, 1905 (Pl. XLVI, Fig. 1–2) was only 30 cm. in amplitude at Fukahori. On the other hand, the seiches between Nagasaki and Fukahori are rather conspicuous at the latter station, where the oscillation forms its loop. Fig. 1 and 2 are two records, when large *abikis* appeared at Nagasaki; Fig. 3 is an example of the record in stormy weather.

VI. BONIN ISLANDS AND FORMOSA.

(1) Futami. Top. 64. Pl. XLVII.

Since December, 1906, a tide-gauge of our system has been set up in the bay of Futami in Bonin Islands (Ogasawarajima) about a thousand kilometers south off the coast of Honshiu. The undulation in the bay is always very regular and conspicuous, with periods **$15.3^m-16.5^m$**, **17.2^m** and **$18.0^m-21.2^m$**, indicating that in this portion of the far ocean, there always exist waves of moderately long periods with greater or less amplitude. The period calculated is 13.6^m which is somewhat less than the observed value.

(2) Kelung. Top. 65. Pl. XLVIII, Fig. 1–2.

Kelung is a harbour at the northern end of Formosa; here Kelvin's tide-gauge has been set up since ten years ago. The undulation is not very conspicuous; the observed periods are **25.3^m–29.6^m** and 57.2^m. The calculated period 25.8^m fairly coincides with the observed.

(3) Takow. Pl. XLVIII, Fig. 3–4.

Takow is situated on the southern coast of Formosa; here Kelvin's tide-gauge is also constantly working. The undulation is very inconspicuous; the periods observed are 11.9^m– 13.7^m and 24.4^m–26.5^m.

§ 5. EXPERIMENTS WITH MODELS.

To confirm our theory, it appeared interesting to experiment with models, and thereby find the actual modes of oscillation of different bays. Models of several bays were formed with definite proportions to the actual dimensions, and the periods of the models were compared with those observed in bays. In reducing the period of the model to the actual one, it was assumed that the period is proportional to length and inversely proportional to the square root of depth, provided the latter is a small fraction of the former.

To construct a model, contour lines of a bay were drawn on separate zinc plates, which were afterwards cut along these lines. The plates were then placed one upon another; the distance between two consecutive plates was kept by blocks of wood of such thickness that it bears a definite ratio to the actual depth. The interspaces between the plates were then filled with cement. The models thus constructed were immersed in a large rectangular tank (150 × 76 × 19 cm.^{3}) filled with water up to the water line of the model.

The waves were excited by a pendulum bob oscillating in

Fig. 6.

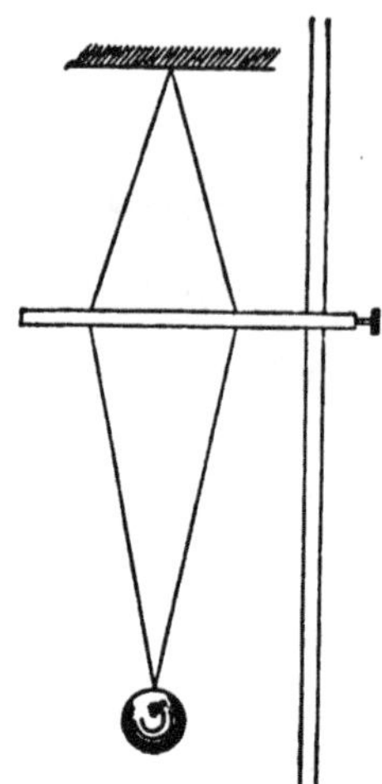

the water; a lead ball. 7 cm in diameter was suspended by two strings passing through two holes in a horizontal rod, as shown in Fig. 6. The part of the pendulum, which oscillated with the bob, was thus restricted to that part of the strings below the horizontal rod; the length of this portion could be varied at will by moving the rod upward or downward. With this arrangement, it was easy to obtain a period less than three seconds. For a longer period, however, it would be necessary to use a pendulum of considerable length. To avoid this inconvenience, a horizontal pendulum was utilized. As shown in Fig. 7, a bar was horizontally supported by means of a string suspended from a knife edge, and of a steel cup, in which the point of the horizontal bar rested. A heavy lead ball was hung by a three-way string from a frame attached to the horizontal rod. By properly inclining the support of the pendulum, periods greater than three seconds could easily be obtained.

Fig. 7.

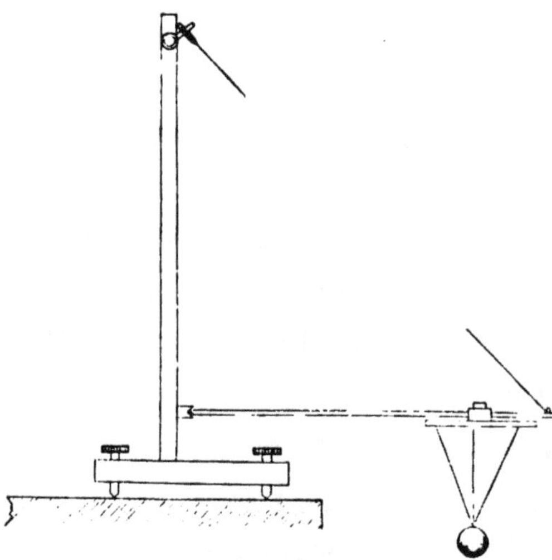

When the pendulum was made to oscillate in front of the model with its bob under the surface of water, the water in the model oscillated smoothly with no appreciable surface wave. Though the amplitude of oscillation of the pendulum gradually decreased, if it were once started and left to itself, it could be kept fairly constant by applying a small force by hand at suitable intervals.

To avoid the reflection of the excited wave from the walls of the tank, a thick layer of a damping material, such as wood shavings, was laid in front of the reflecting walls.

By exciting waves with the above arrangement, the water in the model of the bay made a standing oscillation, whose amplitude was generally small; but as the period of the pendulum approached the proper period of the bay, the amplitude of oscillation gradually increased, and when the period exactly coincided with that of the model the amplitude of the latter was a maximum. In this case, the mode of oscillation proved to be that conceived by us, that is, the end of the bay was a loop for vertical motion and a node for horizontal motion, while its mouth was a node for vertical motion and a loop for horizontal motion. The phase of the water particles in the bay was the same for all parts of the bay, when the oscillation was the fundamental one. In an elongated bay, a binodal or trinodal oscillation was easily produced.

For observing the mode of oscillation of the water, it was convenient to follow the motion of fine cork powders or better fine aluminium powders scattered over the surface of the water. To diminish the effect of surface tension of water on the motion of the powders as much as possible, drops of oil were put into the tank; the powder was then finely scattered, the surface of water being well stirred.

In this way, experiment with the model showed clearly the paths, along which the water particles moved. We also took photographs of the model in the tank, when the bay water was oscillating. By placing a photgraphic camera in a vertical position over the model and exposing the dry plate about half a period of oscillation, a fine path was traced on the plate by each moving aluminium particle, and an aggregate of such tracks of the particles showed beautifully the actual mode of the horizontal motion.

To determine the proper period of oscillation of a model, the period of the pendulum was so adjusted as to give a nearly maximum amplitude of resonance. The pendulum was then stopped, and the period of the subsequent free oscillations was determined by means of a stop-watch. Though the period of the pendulum varied slightly from the proper value, the period of the subsequent oscillation was quite constant.

If the period of the exciter differed considerably from the proper period of the bay, the oscillation after the stopping of the pendulum was rapidly damped, and this gave us a good means of detecting, whether the period of the exciter was near to the proper one, or not.

We experimented with models of seven bays, in which regular and conspicuous undulations were observed ; the results of experiments are given below :—

(*a*) Bay of Hakodate. Pl. LXXXVII.

The scales of the model were as follows :—Length 1 : 20,200, and depth 1 : 548, so that the factor *r*, which was to be multiplied into the observed period in order to obtain the period of oscillation of the actual bay, was 863. The bay had two modes of oscillation ; that is, the one was the fundamental oscillation with its node at the mouth and the other the lateral oscillation between Hakodate and Tomikawa with its node midway between them. The periods of these oscillations were 3.27^s and 1.64^s respectively; multiplying by r, we get 47.0^m and 23.6^m in good coincidence with the observed values. These two modes of oscillation are clearly seen from the photographs No. 1 and 2. Though the photographs show beautifully the line of motion of the water particles, they do not give the direction of motion; hence in Fig. 1 and 2 the direction of motion, as actually observed by experiments, is indicated by arrows. It is very interesting to trace the stream lines in the case of the lateral oscillation (Fig. 2). Certain stream lines extend from Hakodate to Tomikawa and gradually diverge toward the middle, while other lines run toward the mouth of the bay from the Tomikawa side.

In experimenting with models, it was observed that the period of the forcing wave, which corresponds to the maximum resonance, is not well defined; within a certain range of the period, which did not much differ from the period of free oscillation, the oscillation remained fairly conspicuous.

In the actual bay, such a phenomenon was also observed: conspicuous undulations of 45.5^m–57.5^m were frequently observed, the period of free oscillation of the bay being 47.0^m.

(*b*) Bay of Aomori. Pl. LXXXVIII.

The scales of the model were as follows:—Length 1 : 110,700, and depth 1 : 731, so that the facter r was 4,090.

The model had also two modes of oscillation as in the case of Hakodate, i.e. the fundamental and the lateral oscillation. The periods of their free oscillations were 4.45^s and 1.60^s respectively; multiplying by r, we get 303^m and 108^m. During our observations on the actual bay, two oscillations of the periods 300^m and 103^m were observed, which well accord with the above values. The photographs No. 3 and 4, together with Fig. 3 and 4, show the motion of the water particles corresponding to these modes of oscillation. The stream lines (Fig. 3 and 4) in the case of the lateral oscillation deserve a special notice. The greater part of these lines extends from Aomori to Ôminato side, while the remaining lines run from Aomori toward the entrance of the bay; the case is just analogous to the corresponding oscillation in the bay of Hakodate.

(*c*) Bay of Moroiso. Pl. LXXXVIII.

The scales of the model were as follows :—Length 1 : 6,066 and depth 1 : 226 ; the factor r was 404. The fundamental oscillation was 2.28^s; multiplying by r, we get 14.8^m in a good accordance with the observed period. The photograph No. 5 and Fig. 5 show this fundamental mode of oscillation.

(*d*) Bay of Susaki. Pl. LXXXIX.

The scales of the model were as follows :—Length 1 : 24,300, and depth 1 : 915, so that the factor r was 803.

The bay has a very complicated form and many small inlets inside it. By exciting short waves, some inlets energetically oscillated, while the others stood almost still, showing beautifully the phenomenon of resonance.

Waves of different periods were excited and corresponding modes of oscillation of the bay were studied. A period 2.80^s was that of the oscillation of the bay as a whole, period 1.36^s that of the lateral oscillation between Awa and Ôtani, and period 1.25^s that of the oscillation of the small inlet of Kure. Multiplying them by r, we get 37.5^m, 18.2^m and 16.7^m respectively, which agree very well with the observed as well as the calculated values.

The photograph No. 6 and Fig. 6 show a fundamental oscillation of the principal part of the bay as a whole, in which case the small inlet of Kure almost stands still. The photograph No. 7 and Fig. 7 indicate an oscillation of the inlet of Kure, where the water energetically oscillates between Kure and a neighbouring inlet.

(*e*) Bay of Hososhima. Pl. XC.

The scales of the model were as follows:—Length 1 : 10,140, and depth 1 : 366, so that the factor *r* was 531. The fundamental and the binodal oscillations were found to have the periods 2.21^s and 0.90^s; multiplying by *r*, we get 19.6^m and 8.0^m. The phase of the fundamental oscillation is the same for all parts of the bay, while that of the binodal is opposite for the mouth and near the end of the bay; the photographs No. 8 and 9 show also these two modes of oscillation. In the actual bay, the periods corresponding to these two modes of oscillation were also observed.

(*f*) Bay of Nagasaki. Pl. XCI.

The scales of the model were as follows:—Length 1 : 12,130, and depth 1 : 548, so that the factor *r* was 518.

The bay had two modes of oscillation, that is, the fundamental and the seiches-like oscillation. The former had its node at the wide mouth opened to the north-western direction, while the latter had its loops of opposite phases on the Nagasaki and Fukahori sides. The photographs No. 10 and 11 and Fig. 10 and 11, show the two modes of oscillation. It is interesting to observe that in the seiches oscillation (Fig. 11), when the water flows from the wide mouth toward Fukahori, there is also a stream in the direction from Nagasaki to Fukahori, and when the water flows in the opposite direction, there is also a stream directed towards Nagasaki, thus forming an oscillation similar to the seiches between the two places.

The proper periods corresponding to these two modes of oscillation are 2.68^s and 4.45^s; multiplying by r, we get 23.3^m and 38.4^m in good agreement with the observed periods.

(*g*) Bay of San Francisco. Pl. XCII.

During the last fifty years, the tide-gauge at San Francisco has recorded several sea waves which originated at different coasts of the Pacific; the periods of the waves recorded are 17.3^m–19.2^m, 24.3^m–27.8^m, 34.3^m–41.2^m, 47.4^m and 116^m, of which the first is an octave of the third.

Now the Bay of San Francisco is of a very complicated form, so that it is very difficult to find out by calculation what modes of oscillation correspond to the actual periods. Hence a model of the bay was constructed after the chart published by Washington Coast and Geodetic Survey and presented to Professor F. Omori by Dr. O. H. Tittmann, superintendent of the Office. The scales of the model were as follows:—Length 1 : 40,000, and depth 1 : 366, so that the factor r was 2,076. The model was too large to be put in the tank and so it was placed in a small pond of the University. Since the greater part of the model was very shallow, the oscillation rapidly subsided, when the exciting wave was stopped, so that the period was always determined by observing the maximum resonance of the bay. For the exciting wave incident on Golden Gate, the principal modes of oscillation of the water were the oscillations between the West Berkeley and Sausalito sides. The remaining portion of the bay extending to both sides seems to have little influence on these modes of oscillation.

By exciting waves of periods ranging from 3.1^s to 3.5^s, the water in the bay energetically oscillated with the fundamental mode of oscillation, having its node at Golden Gate and its loop at the West Berkeley side. The most easily excitable mode of oscillation was a binodal seiches between the narrowest mouth line and West Berkeley side; the positions of the loops are clearly seen from the photograph No. 12 and Fig. 12 (a chief part of the model). The period of the exciting wave, which gave a marked resonance to the binodal seiches of the bay, ranged from 1.1^s to 1.4^s. By slightly changing the period of the wave, the corresponding displacement of nodal line was observed. We could also produce a trinodal seiches of the bay, whose period of oscillation was 0.8^s. Multiplying these periods by ν, we get 107^m–122^m, 38^m–48^m, and 28.^{m} The period 116^m, which probably corresponds to the fundamental oscillation of the bay, was actually found in the sea wave from South America, 1868, observed in the bay. Periods corresponding to the binodal and trinodal oscillations were often observed in the bay in the case of several sea waves.

§ 6. FORMULA FOR CALCULATING THE PERIODS OF THE UNDULATION IN BAYS.

(*a*) Rectangular bay of constant depth.

The oscillation of the water in a bay is the same as the seiches in a lake of double length, the mouth correction being taken into account. Hence if l and h denote respectively the length and the depth of a rectangular bay of a constant depth, the period T of the free oscillation of the bay, having the node at its mouth and the loop at its end, will be given by the formula

$$T=\frac{4l}{\sqrt{gh}},$$

provided the correction due to the mouth be neglected. This correction may easily be found in the following way.

Take the origin of the rectangular co-ordinates at the middle point on the mouth of the bay, x-axis in the direction of length, positive inwards, and y-axis upwards. Assume the vertical displacement η inside the bay to be given by

$$\eta=a\sin\frac{\pi x}{2l}\cos\frac{2\pi t}{T}.$$

If we neglect the vertical acceleration, we have

$$\eta=-h\frac{\partial\xi}{\partial x}.$$

where ξ is the horizontal displacement of the lipuid element; thus

$$\xi=a\frac{2l}{\pi h}\cos\frac{\pi x}{2l}\cos\frac{2\pi t}{T},$$

and

$$\dot{\xi}=-a\frac{4l}{Th}\cos\frac{\pi x}{2l}\sin\frac{2\pi t}{T}.$$

If b be the breadth of the bay, the kinetic and the potential energy inside the bay are respectively given by

$$\frac{1}{2}\rho hb\int\dot{\xi}^2dx=\frac{4a^2l^3b\rho}{T^2h}\sin^2\frac{2\pi t}{T},$$

and

$$\frac{1}{2}gb\rho\int\eta^2dx=\frac{a^2lbg\rho}{4}\cos^2\frac{2\pi t}{T}.$$

Assume the kinetic energy outside the bay to be

$$Phb^2\rho\dot{\xi}_0^2=\frac{16Pa^2b^2l\rho}{T^2h}\sin^2\frac{2\pi t}{T},$$

where $\dot{\xi}_0$ is the value of $\dot{\xi}$ at $x=o$, and P is of the dimension of a number. Neglecting the potential energy outside the bay, which is very small, we have

$$\frac{4a^2l^3b\rho}{T^2h}\sin^2\frac{2\pi t}{T}+\frac{a^2lbg\rho}{4}\cos^2\frac{2\pi t}{T}+\frac{16\,Pa^2b^2l^2\rho}{T^2h}\sin^2\frac{2\pi t}{T}=\text{const.};$$

the relation is to be satisfied for all values of t.

Putting $t=0$ and also $t=\frac{T}{4}$, we get

$$\frac{a^2blg\rho}{4}=\text{const.},$$

and
$$\frac{4a^2l^3b\rho}{T^2h}+\frac{16\,Pa^2l^2b^2\rho}{T^2h}=\text{const.}=\frac{a^2blg\rho}{4}.$$

Hence
$$T^2=\frac{16\,l^2}{gh}\left(1+4P\frac{b}{l}\right)$$

or
$$T=\frac{4\,l}{\sqrt{gh}}\left(1+4P\frac{b}{l}\right)^{\frac{1}{2}}\ldots\ldots\ldots\ldots\ldots\ldots(2)$$

Lord Rayleigh found the reaction of air upon a vibrating rectangular piston, whose length y is very great compared with its breadth b, to be equal to the addition of a mass

$$y\frac{b^2}{\pi}\left(\frac{3}{2}-\gamma-\log\frac{\kappa b}{2}\right),$$

where γ is Mascheroni's constant and $\doteqdot 0.5772$, and $\kappa=\frac{2\pi}{\lambda}$, λ being the wave length. If the reaction be uniform over the piston, we have for $y=h$,

$$\frac{hb^2}{\pi}\left(\frac{3}{2}-\gamma-\log\frac{\kappa b}{2}\right).$$

Now, in a problem of long waves, we usually neglect vertical acceleration and consider the horizontal acceleration nearly constant for different depths. Vertical planes, which are parallel to wave ridges and fixed relative to water, make a to and fro motion similar to the case of aerial vibration. The nodes of aerial stationary waves correspond to the loop of the water wave and *vice versâ*.* If we use the analogy for the expression of the kinetic energy, we have

$$P \doteqdot \frac{1}{2\pi}\left(\frac{3}{2}-\gamma-\log\frac{\pi b}{\lambda}\right) \doteqdot \frac{1}{2\pi}\left(\frac{3}{2}-\gamma-\log\frac{\pi b}{4l}\right)$$

This relation seems to be sufficient for the estimation of the order of magnitude of the mouth correction. It is to be remarked that if we assume

$$\eta=\sum a_k \sin\frac{k\pi x}{2l}\cos\frac{2\pi k t}{T},$$

the result is not altered.

In the following table, the mouth corrections are given in terms of the ratio of the breadth to the length of the bay :—

$\frac{\text{Breadth}}{\text{Length}}$	$\frac{\text{Corrected Period}}{\text{Uncorrected Period}}$
1	1.340
$\frac{1}{2}$	1.262
$\frac{1}{3}$	1.218
$\frac{1}{4}$	1.187
$\frac{1}{5}$	1.163
$\frac{1}{10}$	1.107
$\frac{1}{20}$	1.064

Here it is to be remarked that the applicability of the above formula becomes less, when the ratio of the breadth to the length becomes greater than unity, since in such a bay, the nonuniformity of horizontal displacement for each transverse section becomes too great.

(*b*) Irregularly shaped bay.

The problem of finding the period of oscillation in an irreg-

*) See Lamb, Hydrodynamics. 3rd Ed. § 188.

ularly shaped bay is also reducible to that of seiches in an irregularly shaped lake. Professor Chrystal* in his hydrodynamical theory of seiches, worked out this problem in a most elegant manner; he compared the oscillation of an irregularly shaped lake with vibration of a string with variable linear density, and gave a minute discussion on a number of special cases.

When the shape of a lake does not much differ from a rectangular tank, the following method of calculating the period may be of some practical importance. In accordance with the above supposition, we may assume that the normal velocity at any section S is constant over it, and that the elevation of the free surface is the same along the entire breadth corresponding to the section S. The vertical acceleration is neglected in comparison with the horizontal.

Take the origin of rectangular co-ordinates at one end of the lake, x-axis being in the direction of the length. ξ, η, b have the same meaning as before. Then the kinetic and the potential energy are respectively given by

$$\text{K.E.} = \frac{1}{2}\int \rho S\dot{\xi}^2 dx \qquad \text{and} \qquad \text{P.E.} = \frac{1}{2}\int gb\rho\eta^2 dx$$

where S is the sectional area.

Again, from the equation of continuity, we have, putting $S\xi = X$,

$$b\eta = -\frac{\partial X}{\partial x},$$

hence $$\text{K.E.} = \frac{1}{2}\rho\int \frac{\dot{X}^2}{S}dx \quad \text{and} \quad \text{P.E.} = \frac{1}{2}g\rho\int \frac{1}{b}\left(\frac{\partial X}{\partial x}\right)^2 dx.$$

---

*) Prof. Chrystal, Trans. Roy. Soc. Edin. Vol. XLI, p. 599, 1905. Prof. Chrystal and E. Maclagen-Wedderburn, do, p. 823, 1905.

If we assume for X an expression of the same form as would be obtained, if S were constant, the length being straight or curved, that is

$$X = \sum a_k \sin \frac{k\pi x}{l} \cos p_k t$$

or

$$= \sum \sin \frac{k\pi x}{l} \cdot \phi_k, \qquad \phi_k = a_k \cos p_k t$$

then

$$\dot{X} = \sum \sin \frac{k\pi x}{l} \dot{\phi}_k,$$

and

$$\frac{\partial X}{\partial x} = \sum \frac{k\pi}{l} \cos \frac{k\pi x}{l} \cdot \phi_k;$$

hence

$$\text{K.E.} = \frac{1}{2}\rho \int \frac{1}{S}\left\{\sum \sin \frac{k\pi x}{l} \cdot \dot{\phi}_k\right\}^2 dx$$

$$= \sum\left\{\frac{1}{2}\rho\int \sin^2 \frac{k\pi x}{l}\, dx\right\}\dot{\phi}_k^2$$

$$+ \sum_k \sum_m \left\{\frac{1}{2}\rho \int \frac{1}{S} \sin \frac{k\pi x}{l} \sin \frac{m\pi x}{l}\, dx\right\} \dot{\phi}_k \dot{\phi}_m,$$

and

$$\text{P.E.} = \frac{1}{2} g\rho \int \frac{1}{b} \left\{\sum \frac{k\pi}{l} \cos \frac{k\pi x}{l}\, \phi_k\right\}^2 dx$$

$$= \sum \left\{\frac{1}{2} \frac{k^2\pi^2 g\rho}{l^2} \int \frac{1}{b} \cos^2 \frac{k\pi x}{l}\, dx\right\} \phi_k^2$$

$$+ \sum_k \sum_m \left\{\frac{1}{2} \frac{km\pi^2 g\rho}{l^2} \int \frac{1}{b} \cos \frac{k\pi x}{l} \cos \frac{m\pi x}{l}\, dx\right\} \phi_k \phi_m,$$

where the summation under the sign $\sum\sum$ must be taken such that $k \neq m$.

Since ϕ's are approximately normal co-ordinates, the quantities under the sign $\sum\sum$ are small, so that for simplicity's sake, we may write

$$\text{K.E.} = \sum (A_k + \delta A_k)\dot{\phi}_k^2 + \sum\sum \delta A_{km} \dot{\phi}_k \dot{\phi}_m,$$

$$\text{P.E.} = \sum (C_k + \delta C_k)\, \phi_k^2 + \sum\sum \delta C_{km} \phi_k \phi_m,$$

where $$A_k+\delta A_k=\frac{1}{2}\rho\int\frac{1}{S}\sin^2\frac{k\pi x}{l}\,dx,$$

$$C_k+\delta C_k=\frac{1}{2}\frac{k^2\pi^2\rho g}{l^2}\int\frac{1}{b}\cos^2\frac{k\pi x}{l}\,dx.$$

Hence, we get*

$$p_k^2=\frac{C_k+\delta C_k}{A_k+\delta A_k}-\sum\frac{(\delta C_{km}-p_k^2\delta A_{km})^2}{A_kA_m(p_m^2-p_k^2)}$$

$$\doteqdot\frac{C_k+\delta C_k}{A_k+\delta A_k},$$

or denoting the period of the k th harmonic by T_k,

$$T_k^2=\left(\frac{2\pi}{p_k}\right)^2=\frac{4l^2}{k^2g}\frac{\int\frac{1}{S}\sin^2\frac{k\pi x}{l}\,dx}{\int\frac{1}{b}\cos^2\frac{k\pi x}{l}\,dx}.$$

Putting $S=S_0+\Delta S$ and $b=b_0+\Delta b$, and neglecting the squares and the products of ΔS and Δb, we get

$$T_k^2=\frac{4l^2}{k^2gh_0}\frac{\int\sin^2\frac{k\pi x}{l}\,dx}{\int\cos^2\frac{k\pi x}{l}\,dx}\left\{1-\frac{\int\frac{\Delta S}{S_0}\sin^2\frac{k\pi x}{l}dx}{\int\sin^2\frac{k\pi x}{l}\,dx}+\frac{\int\frac{\Delta b}{b_0}\cos^2\frac{k\pi x}{l}\,dx}{\int\cos^2\frac{k\pi x}{l}\,dx}\right\},$$

where $h_0=\frac{S_0}{b_0}$. If the integration be effected between the limits 0 and l, we get

$$T_k=\frac{2l}{k\sqrt{gh_0}}\left\{1+\frac{1}{2}\frac{\Delta A}{A_0}-\frac{1}{2}\frac{\Delta v}{v_0}+\frac{1}{2}\int\left(\frac{\Delta S}{v_0}+\frac{\Delta b}{b_0}\right)\cos\frac{2k\pi x}{l}\,dx\right\}\dots.(3)$$

where $$A_0=lb_0,\qquad v_0=lS_0,$$

$$\Delta A=\int\Delta b dx,\quad \Delta v=\int\Delta S dx.$$

The above expression can also be transformed into a simpler form :

*) Rayleigh, Theory of Sound, 2nd. Edition, Vol. I. p. 115.

$$T_k = (T_k)_0 \left\{1 + \frac{1}{2}\int\left(\frac{\Delta S}{v_0} + \frac{\Delta b}{A_0}\right)\cos\frac{2k\pi x}{l}\,dx\right\}\ldots\ldots\ldots\ldots(4)$$

by taking v_0 and A_0 such that Δv and ΔA vanish, and putting

$$(T_k)_0 = \frac{2l}{k\sqrt{gh^0}}.$$

By mechanical integration, we can easily evaluate the values

$$\frac{1}{b_0}\int \Delta b \cos\frac{2\pi x}{l}\,dx \qquad \text{and} \qquad \frac{1}{S_0}\int \Delta S \cos\frac{2\pi x}{l}\,dx,$$

and thus arrive at an expression representing the change of period due to a slight variation in the area and volume of the oscillating liquid. The expression shows that any contraction or expansion at the middle part of the lake prolongs or shortens its natural period respectively, and that a contraction or expansion at the end portion shortens or prolongs it respectively.

To apply the above expression to the case of a bay, we need only to consider a lake whose shape is symmetrical with respect to the vertical plane through the mouth line, and to find the period of the seiches in the lake. This period, if it be corrected for the mouth, is the required period of the oscillation of the bay water.

(*c*) Dumb-bell-shaped bay.

The above assumption does not hold for the case, when a portion of the lake is very much contracted. In such case, however, we may treat the problem in quite a different way. When two basins communicate with each other by a narrow canal, the gravest mode of oscillation takes place such that the levels of the two basins rise and fall respectively. If the breadth of the canal is very narrow compared with the two dimensions of the basins, we may assume that the rise and fall of the level is uniform for each basin, and that in the canal the level is invariable, the motion of the water being chiefly horizontal. Then, denoting the areas of the basins by S and S', the breadth, the depth and the length of the canal by b, h and l respectively, the displacement of water in the canal along its length by ξ, and the vertical displacement of the surface of S and S' by η and η respectively, the kinetic and potential energies are given by

$$K.E.=\frac{\rho bhl}{2}\dot{\xi}^2; \qquad P.E.=\frac{\rho g}{2}(S\eta^2+S'\eta'^2).$$

Again, the correction to the kinetic energy on each end of the canal is nearly

$$\frac{\rho hb^2}{\pi}\left(\frac{3}{2}-\gamma-\log\frac{\pi b}{\lambda}\right)\dot{\xi}^2,$$

in which λ is the wave length, if the basins are infinitely wide, and may be considered nearly equal to four times the length of the basin in the direction of oscillation.

Since $\qquad S\eta=-S'\eta' \qquad$ and $\qquad S\eta=bh\xi,$

we obtain, in the usual manner, for the period of oscillation

$$T=2\pi\sqrt{\frac{Sl}{gbh\left(1+\frac{S}{S'}\right)}\left\{1+\frac{2b}{\pi l}\left(\frac{3}{2}-\gamma-\log\frac{\pi b}{\sqrt{\lambda\lambda'}}\right)\right\}}\quad\ldots\ldots\ldots\ldots(5).$$

In order to test the validity of the result, several tanks were made of zinc plates, two circular basins communicating with each other by a canal of uniform rectangular section. These were filled with water to a suitable depth and set into oscillation; the results of experiment were found to agree well with the above theory.

Special interest is attached to the case, when one of the basins becomes infinitely large; the problem is then reduced to the case of a bay communicating with open sea by a narrow neck. Taking $\lambda=\lambda'=4L$, where L is the length of the bay measured along the probable direction of propagation of waves, we obtain from the above equation,

$$T=2\pi\sqrt{\frac{Sl}{gbh}\left\{1+\frac{2b}{\pi l}\left(0.923+\log\frac{4L}{\pi b}\right)\right\}}. \qquad \ldots\ldots\ldots\ldots\ldots\ldots(6)$$

Another simple case frequently met with is that of a dumb-bell shaped bay communicating with an external wide sea by a narrow mouth or neck at the end of one basin. Using the same notations as before, we get

$$2K.E.=\left\{blh+\frac{2hb_2}{\pi}\left(0.923+\log\frac{\sqrt{\lambda\lambda_0}}{\pi b}\right)\right\}\xi^2$$

$$+\left\{b'l'h'+\frac{2h'b'^2}{\pi}\left(0.923+\log\frac{\sqrt{\lambda\lambda'}}{\pi b}\right)\right\}\xi'^2$$

and
$$2P.E.=g(S\eta^2+S'\eta'^2),$$

where λ_0, λ and λ' may be put equal to mean lengths of the two basins for the first approximation. Since $S\eta=bh\xi-b'h'\xi'$ and $S'\eta'=b'h'\xi'$, we get

$$2K.E.=\frac{\dot{X}^2}{c}+\frac{\dot{X}'^2}{c'}$$

and
$$2P.E.=g\left\{\frac{(X'-X)^2}{S}+\frac{X'^2}{S'}\right\},$$

where
$$bh\xi=X,\quad b'h'\xi'=X',$$

$$\frac{l}{bh}+\frac{2}{\pi h}\left(0.923+\log\frac{\sqrt{\lambda\lambda_0}}{\pi b}\right)=\frac{1}{c},$$

$$\frac{l'}{b'h'}+\frac{2}{\pi h'}\left(0.923+\log\frac{\sqrt{\lambda\lambda'}}{\pi b'}\right)=\frac{1}{c'};$$

whence the equation of motion are at once written down;

$$\frac{\ddot{X}}{c}+g\frac{X-X'}{S}=0$$

$$\frac{\ddot{X}'}{c'}+g\left(\frac{X'-X}{S}+\frac{X'}{S'}\right)=0.$$

Eliminating X', and putting $X=e^{ipt}$, we have

$$\frac{S}{cc'g}p^4-\left(\frac{1}{c}+\frac{1}{c'}+\frac{S}{S'c}\right)p^2+\frac{g}{S'}=0$$

or

$$p^4-\frac{cc'g}{S}\left\{\frac{1}{c'}+\frac{1}{c}\left(1+\frac{S}{S'}\right)\right\}p^2+\frac{cc'}{SS'}g^2=0.$$

If we put $S'=mS$, $c'=nc$, then

$$p=\frac{2\pi}{T}=\sqrt{\frac{gc}{2S}\phi},\ldots\ldots\ldots\ldots\ldots\ldots\ldots\ldots(7)$$

where

$$\phi=1+n+\frac{n}{m}\sqrt{(1+n)^2+\frac{n^2}{m^2}+\frac{2n(n-1)}{m}};$$

whence the value of T can be obtained. If $m=n=1$, $\phi=3\pm\sqrt{5}$.

The case, in which both ends of the dumb-bell shaped bay open to the sea, can also be treated in a similar manner.

§ 7. THE METHOD AND THE RESULT OF CALCULATION FOR THE PERIOD OF OSCILLATION.

In the calculation of the period of oscillation in a bay by the simplest formula, it is necessary to evaluate the length and the mean depth of the bay from the charts; the charts used were those published by the Hydrographical Section of the Naval Department. As the mean depth, we took the ratio of the total volume of the water in a bay to the area of the surface. The length of the bay was measured along a median line drawn parallel to what was considered to be the main stream line.

To find the mean depth of a bay, we began with drawing contours on the chart in which the depths at a number of points in the bay referred to low water springs are given. After drawing as many contours as the case requires, we measure by a planimeter areas bounded by successive pairs of contours. These areas multiplied by the corresponding depths, increased if necessary, by the half range, give the partial volumes of water. Dividing the sum of these partial volumes by the area of the free surface, we get the mean depth.

We calculated by formula (1) the period for all observed bays, and also for those not yet observed, which seem to have the forms specially favourable for oscillation. For several typical bays, we also calculated the correction due to the change of the section as well as that due to the mouth, and compared the corrected values with the observed. The calculation was carried out in the following way.

(a) Bay of Aomori.

According to our investigation, the Bay of Aomori oscillates in two different modes, that is, the lateral and the longitudinal oscillation with the periods 103^m and 295^m respectively. Considering the bay as a rectangular tank, whose length is 55.3 km. and whose mean depth is 36.5 m., we get 97.5^m as the period of the oscillation.

In order to obtain the correction due to the variation of the section, we draw on the chart several lines at suitable positions normal to the line of the length, and then taking the length as abscissa and plotting corresponding breadth as ordinates, we get a breadth-diagram. Draw mean breadth line, at a distance equal to the whole surface of the bay divided by the length. Taking now this line as the new axis of co-ordinates, we can easily draw the diagram for $\Delta b \cos \frac{2\pi x}{l}$; whence by mechanical integration, we get 0,113 as the value of $\frac{1}{A_0}\int \Delta b \cos \frac{2\pi x}{l} dx$. Proceeding in a similar way for the sectional area, the value of $\frac{1}{V_0}\int \Delta S \cos \frac{2\pi x}{l} dx$ is found to be 0.062. Applying these values of corrections to 97.5^m, we finally get 106^m as the period of the lateral oscillation of the bay, in good accordance with the observed value 103^m. In this case, the correction due to the mouth is of course unnecessary.

As for the longer period, the calculation of the period as a rectangular bay gives 213^m; the correction due to the variation of the section, as calculated in the manner above described, is 24.8^m. The correction due to the mouth is 46.5^m, so that the corrected value for the period of longitudinal oscillation is 284^m, which fairly accords with the observed value.

(b) Bay of Ôfunato.

The simple calculation of the period of fundamental oscillation on the assumption that the bay is a rectangular tank, whose depth is 19.0 m. and whose length is 8.1 km., gives 39.5^m. The corrections due to the variation of the section and due to mouth are respectively -10.7^m and $+8.6^m$; the corrected value is therefore 36.4^m. In this case, the correction of the section and that of the mouth have opposite signs, the former a little overweighing the latter.

(c) Bay of Tsuruga.

If the bay is considered as a rectangular bay, whose depth is 28.5 m. and whose length is 15.1 km., the period of oscillation is calculated to be 60.0^{m}; the corrections due to the section and to the mouth are respectively -16.8^{m} and $+15.8^{m}$. The corrected value is therefore 59.0^{m} in a fair accordance with the observed period.

The last two are good examples showing that the correction due to the section and that due to the mouth, nearly cancel with each other. Similar remarks apply for many other bays.

When the mouth of a bay is decidedly contracted, the simplest formula completely fails to give the period of oscillation, in which case formula (6) is to be used. We give here two examples of calculation by formula (6) or (7) and show how the results of calculation accord with the observed values.

(a) Bay of Aomori.

This bay may also be taken as an example of double-bays discussed in p. 66. Referring to the annexed figure which shows the general outline of the bay, the shaded portions were considered as necks connecting external sea and two basins S and S'. Necessary data for the calculation of periods, estimated from the chart are as follows :—

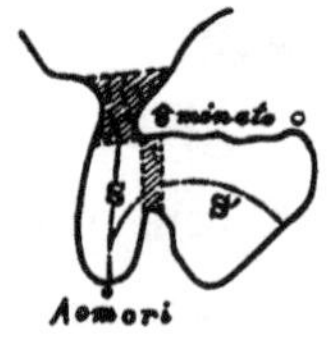

$b=13.5$ km. $b'=13.7$ km.

$l=19.0$ km. $l'=5.3$ km. $\lambda=120$km.

$h=54.8$ m. $h'=41.5$ m.

$S=4.75\times10^{2}$ km.^{2} $S'=9.4\times10^{2}$ km.^{2}

whence for two values of ϕ we obtained 5.25 and 0.49, and for the two periods 100^{m} and $325^{m}_{.}$ in fair accordance with the observed values.

It is interesting to observe that the formula, which is derived from a quite different consideration as compared with the usual view, will give fairly concordant values.

(b) Bay of Ôsaka.

The Bay of Ôsaka which is almost surrounded by land, communicating with external sea by two narrow necks, Akashiseto and Yuraseto, may be taken as an example of those bays discussed in p. 64. For the resultant conductivity of necks, the sum of separate conductivities was duly taken. Data for the calculation of the period estimated from the chart are as follows :

Akashiseto :	$b_1=$ 3.9 km.
	$l_1=$ 6.2 km.
	$h_1=$41. m.
Yuraseto :	$b_2=$ 5.0 km.
	$l_2=$ 1.2 km.
	$h_2=$27. m.
$S=1.47\times10^3$ km^2.,	$\lambda=$150 km.,

whence from (6) $T=270^m$, which fairly accords with the largest period observed.

In the following tables, the mean depth, the length and the period calculated by formula (1) are given in the 4th, the 5th and the 6th columns respectively ; in some bays, the periods corrected for the variation of the section and for the mouth are given with the figures underlined. The observed periods, which would possibly correspond to the calculated values, are given in the last column. The letter S placed after some calculated periods indicates that the periods followed by S are those for the lateral or seiches oscillation in the bays.

Coast of Hokkaidô.

Top. No.	Bay	Province	Depth	Length	Calc. Period	Obs. Period
1	Otaru	Shiribeshi	9.1 m	2.4 km	17.3^m	13.8^m–16.5^m
2	Nemuro	Nemuro	4.2	0.86	9.0	10.9
3	Hanasaki	„	7.5	1.41	10.9	—
4	Bokkiriso	„	5.5	1,7	15.2	—
5	Hamanaka	Kushiro	8.5	6.6	48.2	49.5
6	Akkeshi	„	10.6	13.0	67.2	—
7	Mororan	Iburi	8.1	6.6	48.9	51.1–54.0
8	Hakodate	Oshima	10.7	9.2	45.3	45.5–57.5
			„	9.1	23.6 S.	21.9–24.5

Japan Sea Coast of Honshiu.

Top. No.	Bay	Province	Depth	Length	Calc. Period	Obs. Period
9	Aomori	Mutsu	38.8 m	62.4 km	213;^{m} <u>284</u>^{m}	295.^{m}
			36.5	55.3	97.5, 106 s	103.
10	Yebisu	Sado	135.	10.9	20.1	—
11	Futami	„	16.6	8.4	44.0	—
12	Fushiki	Etchiu	662.	63.4	52.7	54.1
13	Tsuruga	Echizen	28.5	15.1	60.0, <u>59.0</u>	56.7-62.9-67.7
14	Miyazu	Tango	15.5	9.9	53.4	—
15	Tonoura	Iwami	11.6	1.8	11.1	11.9-12.9

Pacific Coast of Honshiu.

Top. No.	Bay	Province	Depth	Length	Calc. Period	Obs. Period
16	Miyako	Rikuchiu	22.0m	10.0km	45.4m	—
			9.7	7.0	24.0 S	21.3m–22.0m
17	Ôdsuchi	"	29.9	7.9	30.7	27.0
18	Ryôishi	"	51.9	7.2	21.3	22.8
			33.1	5.4	20.0	20.3
19	Kamaishi	"	40.8	7.4	24.8	24.8 - 26.0
			19.3	4.6	22.3	20.3
20	Kojirohama	Rikuzen	25.8	6.2	26.0	24.6
21	Yoshihama	"	51.8	7.2	21.1	18.5 - 20.1
22	Okirai	"	31.8	7.0	26.4	27.5 - 29.9
23	Ryôri	"	25.9	4.4	18.4	18.3
24	Ôfunato	"	19.0	8.1	39.5, 36.4	39.0 - 39.1
25	Niiyama	"	11.8	1.2	7.5	6.4 - 7.6
26	Ayukawa	"	12.7	1.5	8.9	6.8 - 8.9
27	Katsuura	Kazusa	7.9	1.8	13.4	—
28	Tateyama	Awa	49.4	6.8	20.5	—
29	Tôkyô	Musashi	52.7	75.5	222.	—
30	Moroiso	Sagami	2.95	1.1	13.4	13.8 - 15.6
31	Sagami	"	461.	19.9	19.8	—
32	Shimoda	Izu	22.2	3.5	15.9	13.8 - 18.2
			15.0	2.4	13.3	—
33	Merakoura	"	12.7	1.4	8.6	—
34	Tago	"	18.8	1.0	4.7	—
35	Heda	"	24.3	1.4	6.1	—
36	Suruga	Suruga	55.7	60.4	54.5	—
37	Shimizu	"	15.6	3.9	21.2	—
38	Ise and Mikawa Sea	Ise and Mikawa	18.4	73.0 (*a*)	363.	390.
			18.4	111. (*b*)	278. S	—
			9.5	31.4 (*c*)	217.	208.

Pacific Coast of Honshiu (*Continued*).

Top. No.	Bay	Province	Depth	Length	Calc. Period	Obs. Period
39	Toba	Shima	18.2m	6.5km	11.4m	—
40	Kukiura	Kii	20.9	3.0	14.1	—
41	Kushimoto	"	15.4	3.4	18.3	16.5m - 18.6m
			5.0	1.3	12.8	11.6 - 13.0
42	Adashika	"	16.4	2.7	14.0	—
43	Ôsaka	Settsu, Izumi and Awaji	—	—	270*	260. - 310.
			27.0	61.0	126 S	106. - 150.
			"	$\frac{1}{3}\times 61.0$	63 S	61. - 66.
44	Hiroshima	Aki	15.0	11.2	61.6	60.0

* The period was calculated by formula (6) as described in p. 71.

Shikoku.

Top. No.	Bay	Province	Depth	Length	Calc. Period	Obs. Period
45	Susaki	Tosa	17.9m	7.8km	39.1m	39.9m-41.6m
	Nomi	"	17.2	6.2	16.0 S†	17.6 - 18.2
	Kure	"	13.2	2.8	17.0	15.0 - 16.3
46	Shimizu	"	7.2	2.9	22.8	—
			"	$\frac{1}{3}\times 2.9$	7.6	—
47	Okuchi	Iyo	29.6	4.5	17.8	—
48	Uwajima	"	204.	16.7	49.8	—
49	Shitama	"	35.0	3.1	11.0	—
50	Yawatahama	"	30.9	4.6	17.6	—
51	Amai	"	24.6	3.3	14.2	—
52	Ikeda	Siôdoshima	9.8	5.5	37.3	—
53	Uchinoumi	"	11.2	7.0	42.9	—
54	Shido	Sanuki	8.4	6.2	45.7	—

† The seiches between Ôtani and Awa.

Kiushiu.

Top. No.	Bay	Province	Depth	Length	Calc. Period	Obs. Period
55	Ariake	Hiûga	42.8m	21.8km	71.1^m	—
56	Hososhima	„	11.7	3.0	19.0	17.8^m–20.3^m
			„	$\frac{1}{3}\times 3.0$	6.3	6.6 – 8.7
57	Yonozu	„	23.3	4.1	18.2	—
58	Aburatsu	„	8.0	2.0	15.1	15.0 – 19.0
59	Kagoshima	Satsuma	103.	51.0	107.	—
60	Nagasaki	Hizen	18.7	7.7	37.5	34.5 – 37.6
			15.3	8.3	22.6 S	22.5 – 25.2
61	Karatsu	„	16.4	13.5	71.0	—
			„	19.7	51.8 S	—
62	Shishimi	Tsushima	13.7	1.5	8.8	—
63	Sasuna	„	11.1	1.9	12.1	—

Bonin Island and Formosa.

Top. No.	Bay	Island	Depth	Length	Calc. Period	Obs. Period
64	Futami	Ogasawara	23.9m	3.1km	13.6^m	16.0^m–20.0^m
65	Kelung	Taiwan	10.4	3.9	25.8	25.3 – 29.6

China and Korea.

Top. No.	Bay	Country	Depth	Length	Calc. Period	Obs. Period
66	Pusan	Korea	10.7m	5.9km	38.4^m	—
67	Onshantin	„	9.9	3.6	30.7	—
68	Weihaiwei	China	8.7	7.4	53.3	—
			„	9.9	33.8 S	—

It will be seen from the above table that in most cases, the period calculated by the simplest formula (1), that is,

$$T = \frac{4l}{\sqrt{gh}}$$

agrees fairly well with those actually observed. Thus for many cases, the corrections due to the mouth and the variation of the section seem to be superfluous. This probably arises from the fact that in many bays, the correction due to the section nearly cancels the mouth correction. For many bays have the form gradually contracting and the depth decreasing, as we approach towards the end, so that the correction due to the variation of the section is negative. The mouth correction being always positive, the two corrections usually tend to annul each other. As exemplified in the cases of Ôfunato and Tsuruga, the total correction nearly vanishes for many bays, and then results an apparent validity of the simplest formula. If the mouth of a bay be contracted and the depth be shallower than in the inside, the correction due to the section is positive, so that the calculated value by the simplest formula decidedly falls short of the observed value, and can only be brought into coincidence by taking the two corrections into consideration ; a good example of this is furnished in the case of the Bay of Aomori. That the calculated periods for the bays of Mororan and Okirai which have rather narrow mouths, are a little less than the observed values, is also explained on the same view.

§ 8. SEA WAVES AND SECONDARY UNDULATIONS.

As we have already remarked, Professor F. Omori found that the periods of sea waves observed in a bay are the same as those of the usual secondary undulation. He explained the phenomenon by supposing that a bay or a certain portion of sea oscillates like a fluid pendulum with its own proper period, when it is excited by an earthquake or any other disturbing causes.

We have also investigated the periods of the sea waves for different bays and found that the above relation is generally well satisfied, especially in the sea waves of distant origin. As we shall see soon below, the period of a sea wave in a bay is, in most cases, given by the formula

$$T=\frac{4l}{\sqrt{gh}}.$$

It is then very probable that in such cases, the sea waves are of a similar nature to the secondary undulations. Now the sea waves are probably of such a complex nature as to be represented by the sum of a series of long waves of different periods and amplitudes. If a group of these waves proceeds towards a bay, the bay takes up and resonates to the undulation, whose period approximately coincides with that given by the above relation.

This consideration chiefly applies to the sea waves of distant origin. If however its origin is not very far from a bay or an open coast, progressive waves of long wave length, irrespectively of their period, are sufficient to cause a disastrous effect on the coast; for by Green's law of amplitudes long waves considerably increase their amplitudes as they approach a shallow shore. Thus in actual destructive sea waves, we almost always have reports of high wave fronts approaching towards the shore, indicating that the waves are of the progressive nature, but not of the stationary character. For the sea waves of 1896, on the coasts of Sanriku, there are instances which seem to indicate that the periods of the sea waves did not coincide with those observed in ordinary cases.

In the investigation of the nature of sea waves, the tide-gauge is the only instrument available at present; but for this purpose, it is necessary to set up the instrument on an open coast, or better on a small isolated island, and not in a bay, where the waves are much modified by the oscillation proper thereto.

In the following pages, discussions regarding to several sea waves in the light of our present investigation will be given.

(*a*) Sea waves of Ansei, 1854.

The destructive sea waves of 1854, which accompanied the great earthquake of Ansei, devastated the great part of our Pacific coast, and were felt by the tide-gauges at San Francisco and San Diego.

On December 23, at $9^h 15^m$ a.m., a strong shock was felt at Shimoda in Izu; at 10^h, it was followed by a large wave 9 m. in height. The rising and falling of the water continued several hours; in all, there were six large waves, the period of which seems to have been 15^m–20^m, if we judge from the descriptions by a sufferer. The period is the same as that usually observed in the bay. A Russian man-of-war then at anchor, was destroyed by the waves.

At Tanabe in Kii, seven or eight waves were observed from the morning to the evening; at Nagashima in the same province, the first wave was the highest, after which two weak waves were counted. At Toba in Shima, the first and fourth waves are said to have been the highest. An irregularity of tides was also observed in Tosa; at Kôchi, three ebb-and-flows were counted from 9^ha.m. to 2^hp.m.

On the 24th, similar waves visited the Pacific coast from Tosa to Izu.

On the 23rd–24th, a regular train of sea waves was recorded on the tide-gauges at San Francisco and San Diego (Pl. XLIX, Fig. 1). The average period of oscillation at San Francisco was 34.9^m, whilst that at San Diego was 34.7^m. The wave took 12^h 12^m and 12^h 37^m to arrive at San Francisco and San Diego respectively.

Now, the Bay of San Francisco, as we have shown by the model, can be put in the oscillation between Sausalito and West Berkeley sides by waves incident on Golden Gate. The period 34.9^m falls a little short of the period of the binodal oscillation of the bay.

The Bay of San Diego is too long to be put in its uninodal longitudinal oscillation; the contraction of the bay near San Diego also prevents the fundamental oscillation. The oscillation between Quarantine wharf and San Diego, which also forms a part of a trinodal oscillation of the whole basin, gives the period of 34.5^m. In this calculation, the depth was estimated from the available charts to be 3.5 m., while its length was taken as 6.07 km.

If the path, by which sea waves travelled from Japan to America across the Pacific, be known, the velocity of propagation can be estimated. Now the sea wave, which has usually a large wave length compared with the depth of the ocean, must be refracted according to the condition of the bottom, so that it is very difficult to know the path by which the sea wave actually travelled through the ocean. We therefore conceived several paths between the origin of the wave to the observed station in question, of which we measured lengths, and also mean depth by mechanical integration.* The velocities of propagation of the long waves along these paths were then calculated from the depths. The time of transmission of the wave along these paths was compared, and the path of the minimum time was taken to correspond to the actual path. From the path thus found and the actual time of transmission, we found the mean velocity of propagation of the sea wave, as given in the following table. As the time of occurrence of the sea wave in its origin, we took the time of the earthquake.

Station	Distance	Mean Depth	$\sqrt{gh}$	Time Interval	Velocity
San Francisco	8,190 km.	5.50 km.	232 $\frac{m.}{sec.}$	12.20^h	186 $\frac{m.}{sec.}$
San Diego	9,000	"	"	12.62	198

Hitherto it has been customary to measure the path along the great circle of the earth; but the actual distribution of the depth being complicated, this is not a proper method.

The value $\sqrt{gh}$ of the fourth column in the above table represents the theoretical velocity of a long wave. Strictly speaking, the mean velocity should be deduced from the distance s along the path and the time interval t† given by

$$t = \int \frac{ds}{\sqrt{gh}}$$

*) The chart published by "Deutschen Seewarte," 1896, was used. One of the authors evaluated the mean depths by Berghaus' Physikalischen Atlas and obtained considerably large values (Proc. Tokyo Math.-Phys. Soc. **3**, p. 165.).

†) C. Davison, Phil. Mag. **50**, p. 579, 1900; H. Nagaoka, Proc. Tôkyô Math.-Phys. Soc., p. 126, 1902.

In deducing the above formula, no account is taken of the curvature of the earth; if this is taken into account, the general tendency is to reduce the velocity from that given by $\sqrt{gh}$. Of course there is some dependence on the wave length and the configuration of the sea. H. N.

where ds is the elementary distance, and h the depth at the point under consideration. It was therefore recalculated by the above relation; but we did not find the increase of its value more than 1 per cent. In the other sea waves, the difference did not exceed the same limit. At any rate, the calculated velocity is considerably greater than the actual value; this point has been noticed by several earlier writers, such as Milne, W. J. H. Wharton, E. Geinitz, C. Davison, etc.

(*b*) Sea waves of South America, 1868 and 1877.

At about 5^h p.m. on August 13, 1868, a destructive sea wave, which followed a severe shock, swept away many cities on the coast of South America. It originated between Iquique and Arica, and was felt at different coasts of the Pacific.

In Northern America, it markedly disturbed the tide-gauges at Astoria, San Francisco and San Diego (Pl. L).* The time of arrival in local time (8^h W.) and the periods observed are given in the following table:—

Station	Beginning	Period
Astoria	$8^h 32^m$ a.m.	24.0^m, 27.6^m, 36.3^m
San Francisco	$2^h 24^m$ a.m.	19.2^m, 36.4^m, 41.2^m
San Diego	$1^h 17^m$ a.m.	31.0^m, 35.1^m, 46.8^m

Here it is to be noticed that the observed periods did not remain constant, but varied within a certain range, the periods

*) Dr. Tittmann presented to Professor F. Omori at his request several valuable records concerning sea waves. The curves in Pl. L are those reduced by a pantograph from the records. Periods given in the above table were obtained from the same records. The same remarks also apply to sea waves of Iquique.

in the above table being the mean values of several oscillations in comparatively regular trains and therefore the period of single undulation may differ from the mean by a few minutes. At San Francisco, the waves of about 41^m appeared most frequently, though the period 35^m of Shimoda wave also manifested itself. By referring to the result of experiment with models, it may be noted that these periods probably correspond to the higher modes of oscillation of the bay. At San Diego, the period 31^m–35^m appeared most conspicuously, which probably corresponds to trinodal seiches of the bay. It is a characteristic feature of this sea wave that at these stations, the initial waves are not conspicuous and their amplitudes gradually increase.

The mean velocity of propagation of the wave from South America to Astoria, San Francisco or San Diego can be calculated, if the path through which the wave actually travelled be known; but this is not an easy matter, the condition of the bottom along the coast line of America being complex.

In Japan, the wave was observed at Hakodate by Captain T. Blakiston, who wrote the following passage to Professor Milne :—

"On August 15, at $10^h 30^m$ a.m., a series of bores or tidal waves commenced, and lasted until 3^h p.m. In ten minutes, there was a difference in the sea level of 10 feet, the water rising above high water and falling below low water mark with greater rapidity."

Now the periods of the ordinary conspicuous undulation in the Bay of Hakodate are 45^m–57^m and 22^m–24^m. In the case of sea waves, the latter period always appears superposed on the former. According to the report of the captain, half the period of the present sea wave is 10^m; allowing errors of 1 or 2 minutes of observation, it is probable that the wave corresponds to the oscillation proper to the bay with the shorter period, probably superposed on the longer period. The wave took $24^h\ 57^m$ to arrive at Hakodate across the Pacific.

As in the former case, we calculated the mean velocity of the sea wave with the result given in the following table:—

Station	Distance	Mean Depth	$\sqrt{gh}$	Time Interval	Velocity
Hakodate	16,600 km.	4.78 km.	2.14 $\frac{m}{sec.}$	24.95^h	185 $\frac{m.}{sec.}$

Thus the mean velocity of the sea wave fairly coincides with that between Shimoda and San Francisco; it is decidedly less than the value of $\sqrt{gh}$

The terrible earthquake of Iquique on May 9, 1877, also caused destructive sea waves, which were felt across the basin of the whole Pacific, from New Zealand in the south to Japan and Kamschatka in the north. In Japan, the waves were observed at Hakodate, Kamaishi and the coast of Kazusa.

The earthquake originated at $8^h 20^m$ p.m. on the 9th, and the consequent sea waves broke on the shore of Iquique 30^m afterwards with disastrous effect. At $6^h 21^m$ a.m. on the 10th, a group of sea waves suddenly invaded the bay of San Francisco (Pl. XLIX, Fig. 2) and put the bay water into oscillation; the oscillation seems to have been renewed at $2^h 50^m$ p.m. by another group of incident waves. The oscillations continued over 2 whole days; their periods of oscillation in comparatively regular trains were 17.3^m, 27.8^m, 34.3^m and 47.4^m. The most conspicuous periods were 34.3^m and its octave 17.3^m, the former falling near the period of the binodal oscillation between Sausalito and West Berkeley sides.

In Japan, at $11^h 30^m$ a.m. on the 11th (in our time), the sea at Hakodate was observed to rapidly retire, and then to rise in level to about 2 m.; the rising and falling of the level lasted the whole afternoon with a period of about 20^m. Between $2^h 30^m$ and $2^h 35^m$ p.m., the oscillation was renewed with the largest amplitude of 2.4 m. Admitting errors of observation of a few minutes, the oscillation possibly corresponds to the oscillations proper to the bay.

The tide in the bay of Kamaishi began to oscillate between 9^h and 10^h a.m. of the same day, and the amplitude of oscillation gradually diminished. At 0^h and 2^h p.m., the phenomenon was renewed; and until 5^h or 6^h p.m., the bay was observed to oscillate with an amplitude of 3 m. and with period of 5^m. At midnight, the sea was completely calm. Comparing the waves with those in the Bay of Hakodate, we notice that the first wave was not observed in the latter bay.

At noon of the same day, large waves invaded the open coast of Kazusa, but the sea soon became calm. At 4^h p.m., still larger waves devastated the same coast, causing the loss of many lives. These two waves possibly correspond to the waves which visited the bays of Hakodate and Kamaishi a little earlier.

The calculation of the velocity of propagation of these waves gave the following results :—

Station	Distance	Mean Depth	$\sqrt{gh}$	Time Interval	Velocity
Hakodate	16,600 km.	4.78 km.	214 $\frac{m.}{sec.}$	25.00^h	185 $\frac{m.}{sec.}$
Kamaishi	,,	4.80	216	22.92	201
Kazusa	,,	,,	,,	25.25	183

Thus, the velocities in the first and third rows in the above table are decidedly less than the velocity in the second row; this discrepancy is probably due to the fact that, the first wave was not observed at Hakodate and Kazusa probably on account of its small amplitude. Here also the value of $\sqrt{gh}$ is a little greater than the actual velocity.

(*c*) Sea waves of the Krakatoa eruption, 1883.

Great sea waves caused by the eruption of Krakatoa, August 27, 1883, swept over the entire area of the Indian ocean and forced their way as far as to the northern parts of the Atlantic and the Pacific, leaving their traces on the tide-gauges situated along their way, the records of which are reproduced in the Report of the Royal Society of 1888, and described by Captain W. J. H. Wharton. These records as well as numerous matters of information from different quarters of the world, in which the Report abounds, if reviewed under the light of the results of our present investigations, may be of some interest, inasmuch as the former reporter seems to have put a little weight on the secondary undulation peculiar to bays and estuaries.

As to the cause of the periods of the great Krakatoa waves, few theories have been proposed. Captain Wharton* attempted

*) Wharton, The report of the Krakatoa eruption, p. 97.

to explain the period of two hours by assuming that the sea bottom was upheaved for about an hour. According to Professor H. Nagaoka,* the earth is continuously vibrating with a period 67^m of fundamental oscillation, and it is this vibration that actually determined the periods of the Krakatoa waves. Our theory differs from the above by not assuming the slow up and down motion of sea bottom. Now, according to the results of our observations, any portion of sea partly bound by land, is capable of its own mode of stationary oscillation, which if properly excited, may last for some time, after the cause of the excitement has receded. In this respect, the Sunda Strait (Top. 69) presents a highly suggestive form by its boundary. The south-west end of the channel opens widely into the Indian ocean, while the north east end is narrow and shallow communicating to the Java Sea. The strait, as a whole, may be compared in its acoustical analogy with a conical open pipe, provided that in the case of the hydrodynamical problem, both ends are to be considered as the nodes of the wave profile. The loop of the gravest mode of oscillation possible in such a channel must lie nearly midway but somewhat nearer the narrower end. Hence the eruption of Krakatoa lying nearly at the loop of this oscillation would be very favourable to excite the natural stationary oscillation of the strait as a whole. The initial disturbance, would soon settle into a regular oscillation natural to the system; this again would be propagated into the external ocean as a train of regular waves, whose period of oscillation is determined by that of the source.

Taking the length of the strait as 160 km., and its mean depth as 183 m. (=100 fathoms), we obtain from our formula

*) Prof. H. Nagaoka, Proc. Tôkyô Math. Phys. Soc. 4. p. 35, 1907; Nature, May 24, 1907.

$T=126^{m}$, which was actually recorded in the tide-gauge of Batavia (Pl. XLIX, Fig. 3)*. Though the position of the node at the southwest end is not very determinate, the ambiguity does not affect the value of the calculated period in any serious manner, because the bed to this end slopes down very steeply toward the deep sea, so that if the virtual length of the strait be assumed to be a little longer or shorter, the mean depth of the oscillating basin becomes greater or less in a considerable proportion, so that the value of $l/\sqrt{h}$ remains fairly constant. The fact that at Anjer at the northwest mouth of the strait, the reported height of the waves was small, seems to stand in harmony with our supposition that the node lies near that place.

Besides the mode of oscillation above described, a binodal oscillation between the two sides of the strait, Java and Sumatra sides, might possibly be generated by the eruption of Krakatoa, which lies at the loop of this mode of oscillation. The period of the oscillation is calculated to be about one hour, which nearly coincides with the periods recorded at many stations along the Indian coast. In addition, the higher modes of oscillation than the above two with comparatively small amplitudes might possibly have been in co-existence.

Beyond the northeast end of the strait, the sea is shallow and abounds in irregularities of bed, which may scatter the waves propagated from the end of the strait by complicated reflection and refraction. Besides, the sectional area of the northeast end of the strait is estimated to be about $\frac{1}{60}$ that of the southwest end, so that the energy propagated from the former mouth must have been a small fraction of that from the

*) All tide-gauge records regarding the Krakatoa eruption are the reproduction from the Report of the Royal Society of London.

latter. These considerations probably account for the smallness of waves propagated in the northeastern direction.

The tide-gauge nearest Krakatoa at the time of the eruption was that of Batavia. It beautifully recorded two hour waves, but did not trace one hour waves. The absence of this latter period in Batavia raises no serious objection against our supposition, because the narrow opening to the northeast of the strait, is very unfavourable for the propagation of the energy of the lateral oscillation, as compared with that of the longitudinal.

Thus the energy of the oscillations was, in its greater part, propagated into the Indian Ocean (Pl. LI) and strikingly affected the tide-gauges so far as the ports of Southern Africa (Pl. LII, Fig. 1–2). Examining the records given in the above cited reports, we may generally distinguish two types of waves,—one includes those types of waves propagated directly from Krakatoa and the other the stationary oscillation of bays or estuaries excited by the incident waves. Prominent undulations recorded along the coast of India belong to the former type. Comparing the records at Madras and Vizagapatum or Negapatum and Port Blair, the identity of waves may easily be recognized. We see also the trace of Vizagapatum waves in the Negapatum record and *vice versâ*. Most of these stations are not situated in either bay or estuary possible of oscillation with such a long period.

For remoter stations, we see generally that the disturbances are chiefly due to the second type, i.e., to the proper oscillation excited by the synchronizing components of the incident waves. Hence for such bays, the periods of oscillation for Krakatoa waves may be calculated from their dimensions, provided a good chart is at hand. Since, at present, we are in want of reliable chart on sufficient scales, we satisfy ourselves by deducing the mean depths of different bays, quoted in the Krakatoa Report, from their period of oscillations and their lengths estimated from the charts* available. The results of calculation are as follows :—

(1) Port Elizabeth, Cape Colony. Pl. LII, Fig. 1.

The node of the Algoa Bay was taken from Cape Receife to the coast of Alexandria opposite to the Bird Island; taking the length $l=25$ km. and the period $T=74^m$, we found the mean depth h to be 52 m.

(2) Table Bay, do. Pl. LII, Fig. 2.

The node was taken from Green Promontory to the coast opposite to Rabben Island; taking $l=11.3$ km. and $T=60^m$, we get $h=16$ m.

It will be noticed that the shape of the above two bays is very similar to that of Hakodate, and also that the records of secondary undulations in these bays are, in their character, quite similar to each other.

(3) Port Adelaide, Australia. Pl. LII, Fig. 3.

The period of 150^m traceable in the record of Port Adelaide may be explained, if we suppose it due to the oscillation between Port Adelaide and York Peninsula. In this case we obtain $h=21.5$ m. which very nearly coincides with the result of our estimation from the chart. Besides, a period of about 200^m is traced, which is probably due to the oscillation of the Investigator Strait, with nodal lines at its two ends. The period calculated for the latter modes is 228^m.

(4) Port Phillip. Pl. LII, Fig. 4.

---

*) Berghaus' Physikalische Atlas, Stieler's Hand Atlas, Encyclopedia Britanica etc.

The period of 86^m may be accounted for, if it be attributed to the binodal seiches of the enclosed basin.

(5) Lyttelton, New Zealand. Pl. LIII, Fig. 1.

Very conspicuous oscillation of this port with the unusually long period of 165^m is probably due to the Pegasus Bay. Taking the node between Table Island and the end of Bank Peninsula, the calculated mean depth is about 30m. which seems allowable. On the other hand, if we suppose the period due to the oscillation in the narrow inlet of Lyttelton, the calculated mean depth would only be 8 m.

(6) Honolulu, Hawai. Pl. LIII, Fig. 2.

Taking $l=2.25$ km. for the narrow inlet and $T'=27.7^m$, we obtain $h=3.7$m, which seems reasonable. Another conspicuous period is probably due to the binodal oscillation of the inlet.

(7) San Francisco. Pl. LIII, Fig. 3.

The conspicuous periods of the Krakatoa waves observed in the bay were 24.0^m, 36.2^m and 48.6^m. These periods were frequently found in the same bay for other sea waves, and as we have already remarked, probably correspond to the multinodal seiches between the West Berkeley and Sausalito sides.

(*d*) Sea wave of Sanriku, 1896.

On June 15, 1896, a destructive sea wave originated in a distance of about 150 km. off the coast of Sanriku in Japan. The wave was the most disastrous one in modern time; its height in Yoshihama even amounted to 24 m. It swept away many towns and villages along the coast line of Sanriku, extending to about 320 km.; 22,000 lives were lost.

At Miyako in Rikuchiu, the earthquake was felt at $7^h 32^m$ p.m. and the tide began to retire at about $7^h 50^m$; it then increased and attained a maximum at about 8^h. Then it decreased and again increased; at $8^h 7^m$, the largest wave invaded Miyako. The subsequent large waves rolled on the shore at the following epoch:—

$8^h 15^m$, $8^h.32^m$, $8^h.48^m$, $8^h.59^m$, $9^h 16^m$, $9^h.50^m$;

the intervals between two consecutive waves are 8^m, 17^m, 16^m, 11^m, 17^m and $34.^m$ The period of the conspicuous undulation of the bay of Miyako commonly observable in ordinary weather is 21^m, which is somewhat different from the period of the present sea wave. From the above intervals, it may be concluded that the sea wave was probably composed of the period of about 16^m and its octave. Without the aid of a tide-gauge, it would often be difficult to observe every incident wave of 8^m out of waves so composed.

The record of the nearest tide-gauge was that of Ayukawa (Pl. LIV, Fig. 1); it shows a series of gigantic waves of the period 8^m, which continued over two days; during the first twelve hours, the rising and falling of the level followed each other most energetically. Comparing the period of the wave with that observed at Miyako, it may be concluded that the period of the waves incident on the bay nearly coincided with the period of its free oscillation. Examining the record of the tide-gauge, we observe, in the latter part of it, a beautiful series of beats which probably shows that the period of the incident wave is slightly different from that of the free oscillation of the bay.

The sea wave also slightly affected the tide-gauge of Aburatsubo in Misaki (Pl. LIV, Fig. 2), which is about 600 km. distant from its origin. Its period of undulation was 15^m in a good coincidence with that usually observable. The same wave also affected the tide-gauge at Hakodate; unfortunately, this instrument stopped at the very beginning of the sea wave, and began to work again after several hours. The oscillation of the bay continued over two days with its proper periods of 23.6^m and 45.5^m–57.5^m (Pl, III, Fig. 2).

In the above four cases, the wave was more or less affected by the proper oscillations of the bays. The tide-gauge at Chôshi in Shimôsa has however been set up in a mouth of the river Toné, so that its record in the case of the sea wave is the most suitable for the investigation of the wave, inasmuch as the sea is not much affected by any proper oscillation of enclosed water. The record of the tide-gauge at Hanasaki, which is situated on a small inlet in the Pacific coast of Nemuro, possesses the same advantage as that of Chôshi. Thus, in the records of Hanasaki and Chôshi (Pl. LV, Fig. 1–2), we observe, for the first one hour,* a similar series of waves of the period of about 7^m. This period was found superposed on the larger waves of some ten minutes.

The sea wave also crossed the Pacific and reached the western coast of America. It disturbed the tide-gauges of Honolulu and San Francisco (Pl. LV, Fig. 3–4). The period of the wave in Honolulu was 23.4^m– 26.0^m which is nearly the same as that of Krakatoa waves ; in the first part of the record, we may however observe a wave of double period. The record of San Francisco was marked by irregular zigzags ; we can however trace waves of periods 24.3^m and 6.2^m

In the following table, the result of the calculation for the velocity of sea waves through the Pacific are given :—

---

*) After the first hour, the tide-gauge at Hanasaki stopped.

Station	Distance	Mean Depth	$\sqrt{gh}$	Time Interval	Velocity
Honolulu	6,000km.	4.92 km.	220 $\frac{m.}{sec.}$	$7^h 44^m$	216 $\frac{m.}{sec.}$
San Francisco	7,970	5.51	234	$10^h 34^m$	209

Thus the mean velocity of propagation of the sea wave is somewhat greater than the velocities of other waves. The mean value of $\sqrt{gh}$ between Sanriku and Honolulu fairly agrees with the observed, while the value between Sanriku and San Francisco is decidedly greater.

(*e*) Sea waves of South America, 1906.

The earthquake of Ecuador originated, in our time, at $0^h 42^m$ a.m., February 1, 1906, and was accompanied by a sea wave, which traversed across the Pacific from America to Japan. The wave was recorded by the tide-gauges of Hakodate, Ayukawa, Kushimoto, Hososhima and Fukahori (Pl. LVI, Fig. 1–4; Pl. LVII, Fig. 1). The earthquake of Valparaiso, which was also followed by a sea wave, originated at $9^h 45^m$ a.m., July 17 in the same year. The wave arrived at Hakodate, Ayukawa, Aburatsubo and Kushimoto (Pl. LVII, Fig. 2–3; Pl. LVIII, Fig. 1–2). In the two waves, the periods observed in these bays were as follows:—

Stations	Ecuador Wave	Valparaiso Wave	Ordinary Case
Hakodate	49.2^m–40.9^m	53.0^m–48.0^m	57.5^m–45.5^m
	21.9^m	22.1^m	24.5^m–21.9^m
Ayukawa	7.3^m–8.3^m, 20^m	7.3^m, 23^m	6.8^m–8.9^m, 20.9^m–22.8^m
Aburatsubo	—	96.0^m	13.8^m–15.6^m
Kushimoto	21.0^m, 13.2^m	21.4^m, 12.3^m	21.5^m–23.7^m, 11.6^m–13.0^m
Hososhima	20.4^m	—	17.8^m–20.3^m
Fukahori	10.4^m, 13.6^m	—	12^m

Thus in these bays the periods of the sea waves are the same as those observed in other sea waves and also in ordinary cases. Only one exception is found in the bay of Aburatsubo for Valparaiso wave the period of which is several times greater than the period of free oscillation of the bay.

The following tables contain the results of calculation for the velocity of the sea waves through the Pacific.

Ecuador Wave.

Station	Distance	Mean Depth	$\sqrt{gh}$	Time Interval	Velocity
Hakodate	14,330km.	4.92 km.	220 $\frac{m.}{sec.}$	$20^h 16^m$	195 $\frac{m.}{sec.}$
Ayukawa	"	"	"	$20^h 12^m$	200
Kushimoto	15,280	4.81	217	$20^h 44^m$	208
Hososhima	15,610	"	"	$20^h 38^m$	211
Fukahori	16,080	"	"	$21^h 5^m$	213

Valparaiso Wave.

Station	Distance	Mean Depth	$\sqrt{gh}$	Time Interval	Velocity
Hakodate	17,080km.	4.66 km.	214 $\frac{m.}{sec.}$	$23^h 48^m$	200 $\frac{m.}{sec.}$
Ayukawa	"	"	"	$23^h 17^m$	204
Aburatsubo	17,280	"	"	$22^h 2^m$	217
Kushimoto	17,600	4.58	212	$23^h 31^m$	208

From the above tables, it is easily seen that notwithstanding the considerable distance travelled by these sea waves, the velocities in the two cases fairly coincide with each other. In comparing velocities of the same wave at different stations, it will be observed that the velocity slightly increase with distance, though the mean depth of the ocean remains nearly constant or rather decreases slightly. The exceptionally large velocity in the case of Aburatsubo, when sea wave had a considerable long period, is probably due to the effect of the wave length, on the velocity. The increase of velocity with distance is also noticed in the case of the sea waves accompanied by the Krakatoa eruption. This seems to indicate that the earthquake may excite short waves almost simultaneously with shocks, but it takes some time to excite long waves of a permanent type of a considerable period.

In the sea waves above referred to, the observed velocities nearly coincide with the values of $\sqrt{gh}$, except for Hakodate and Ayukawa, in which the difference is considerable.

The sea waves also slightly affected the tide-gauge* at San Francisco, San Diego and Honolulu (Pl. LVIII, Fig. 3–4; Pl. LIX, Fig. 1–3). The times of arrival of the waves in the standard times of these stations and the periods of the waves are given in the following table :—

	Ecudor Waves		Valparaiso Waves	
Stations	Beginning	Periods	Beginning	Periods
San Francisco	—	—	$7^h 34^m$ a.m.	25.3^m
San Diego	$4^h 45^m$ p.m.	16.7^m, 33.5^m	$6^h 25^m$ a.m.	32.2^m, 43^m
Honolulu	$8^h 25^m$ p.m.	24.8^m, 26.8^m	$7^h 52^m$ a.m.	26.2^m

Thus, if we compare the periods of several sea waves observed at these American bays, it may also be concluded that the periods of the different sea waves observed at each bay are proper to the bay. Honolulu affords the best example; a simple wave of the period varying from 23^m to 28^m has always been found for sea waves originated near the Pacific coast of America

*) The tide-gauge records utilized are those furnished by Dr. Tittmann.

or of Japan. As we have seen, this period nearly coincides with that of the proper fundamental oscillation of the inlet leading to Honolulu.

(*f*) Sea waves accompanying cyclonic storms.

Hitherto we have exclusively considered the sea waves caused by the earthquake or the submarine eruption. There are, however, other classes of sea waves accompanying a cyclonic storm. These latter may be subdivided into three kinds, that is, short and long waves and abnormal rise of sea level.

The violent short waves, the *gekirô* as they are commonly called, have usually periods of a few minutes, and are superposed by waves of still shorter periods with considerable amplitudes. On the Pacific coast they cause, year after year, great damages in autumn, and on the Japan Sea coast, in winter. The waves are always associated with strong gales; as they approach a shallow shore, they increase in amplitudes, and break on the shore one after another, leaving damage behind them. They have probably the same origin as the ordinary wind waves* always observable on the surface of the sea.

Examining the records of the tide-gauges, it will be observed that there are many cases, in which general sea-level is abnormally raised for many hours, associated with a center of low pressure existing near by. For example, the record of the tide-gauge in Aburatsubo (Pl. LX, Fig. 1), September 28, 1902, shows a remarkable upheaval of level in 7^h–10^h a.m. during low water, the maximum height above the ordinary level being about 20 cm. On examining the weather chart on the same day (Pl.

*) Wheeler, Tides and Waves, Ch. X, Wind waves.

XCIV), it was found that a remarkable cyclonic center drew across the district from the Pacific to the Japan Sea during these hours. This general upheaval is also accompanied by the secondary undulation proper to the bay. To take another example, the mareogram in Takow, Formosa (Pl. LX, Fig. 2), June 27, 1904, shows an abnormal rise of sea level; at 4^h a.m., the sea began to rise rapidly, attained a height of about 60 cm. above the ordinary level after half an hour and then gradually returned to its ordinary level. The weather chart (Pl. XCIV) shows a persistent center of low pressure on the south of Formosa during preceding days accompanied by lasting eastern gale in that district, which seems to have attained its height on the morning of 27th. This abnormal rise of the level is probably due to the strong gale.

There are further marked sea waves of the same kind, which are however not recorded on the tide-gauges. On August 11, 1889, a deep cyclonic center rapidly approached from Kii sea to the Bay of Mikawa and reached that district with a high velocity. When the center was passing over the bay, strong gales upheaved the sea level by a few meters for an interval of about 1 hour. The short waves, superposed on this abnormal rise of the level, broke on the coast, leaving great catastrophes behind them. On August 28, 1902, a cyclonic center was threatening the southern coast of Tôkaidô. At the same time, another center of low pressure, appearing in the vicinity of the Bonin Island, drew towards the Bay of Sagami and crossed over the central part of Honshiu. When the center arrived at the coast, strong gales heaped up the water and flooded the shore of Odawara and its vicinity. Associated with the upheaval of the level, the short waves rolling into the shore caused great damage to the town.*

There are a few examples, in which abnormal rise of sea level occurred, when neither a cyclone nor an earthquake was recorded. For example, on March 4, 1901, the mareogram at Hososhima (Pl. LX, Fig. 3) shows an upheaval of about 20 cm. during 7^h–10^h a.m. In the weather chart on this day (Pl. XCIV–XCV), no cyclone is indicated, though a low pressure seems to have prevailed over the Pacific and gales are recorded in several other stations in southern Japan. On this occasion, the secondary undulation was unusually faint.

As for the remaining kind of sea waves,† namely long waves with considerable amplitude accompanying cyclonic storms, we have many remarkable examples, some of which we shall describe. On the morning of September 28, 1900, a center of intense cyclonic disturbance passed over the southern coast of Kii (Pl. XCV). The tide-gauge record at Kushimoto (Pl. LXI. Fig. 1), the southernmost promontory of Kii, shows a rise of general level amounting to about 40 cm. from 3^h to 6^h a.m., upon which very remarkable long wave with a maximum amplitude of 80 cm. were superposed. The period of the most conspicuous wave is about 10^m. It is a remarkable fact that on this occasion, the long wave with considerable amplitude lasted only for 3 or 4 hours, during which the general level was upheaved; they soon subsided into ordinary small waves with the rapidly disappearing cyclonic center. A similar case occurred on August 28, 1899, though without considerable upheaval of the general level. To take another example, on the

---

*) Many examples of the upheaval of sea level observed in Europe and America will be found in Wheeler's "Tid s and waves," Chapter VIII.

†) Wheeler, "Tides and waves," Chapter XI, p. 137–140.

afternoon of September 28, 1902, a remarkable cyclonic center crossed over the northern part of Honshiu (Pl. XCIV) and the tide-gauge at Ayukawa showed remarkable secondary undulation with the period of about 7^m (Pl. LXI, Fig. 3), which lasted for more than 12 hours. The maximum amplitude recorded was about 1 m., nearly at the time of arrival of the cyclonic center. It is to be remarked that in these examples the periods of long waves with remarkable amplitudes are nearly equal to those peculiar to the bays.

In the above examples, the beginning of the remarkable undulation is well defined and its connection with the presence of cyclonic center is very clear. On the other hand, there are examples in which remarkable secondary undulations lasted for a very long time during cyclones passing along our land, the beginning being not quite marked. Generally speaking, the duration of large secondary undulations seems to depend on the width as well as the velocity of the cyclonic area.

There are also cases, in which a remarkable undulation appeared, when neither a cyclone nor an earthquake was reported. As we have already remarked, the mareograms of Nagasaki afford many good examples of this sort of undulation. To take another example, the Kushimoto mareorgram (Pl. LXII, Fig. 1) on February 8, 1904, is marked with a considerable undulation with the periods of 13^m, 20^m–26^m, etc., which lasted from the morning till the next day, the greatest amplitude being nearly 60 cm. The weather chart indicates no cyclone and no wind in the neighbourhood of the district, but low pressures reigned over the northern part of Hokkaidô and also over the Pacific in the south of Honshiu (Pl. XCV), while high pressure prevailed over the continent. The remarkable undulation seems to be associated with unstable distribution of pressure, in which case a sudden local change of pressure may be possible.

It is to be remarked that at a station, the period of the most prominent undulations accompanying a cyclonic storm are sometimes different for different cases. Thus, in Kushimoto, the most prominent undulations of the sea wave of February 8, 1904, were of 21^m–24^m, while the period recorded at the time of cyclone on September 28, 1900, was about 10^m. Again, during the cyclone on November 17, 1900, the Ayukawa tide-gauge (Pl. LXII, Fig. 2) showed a remarkable undulation with the period of 22^m, superposed by waves of shorter period peculiar to the bay, while in the cyclone of September 28, 1902, the proper period of the bay are most pronounced.

Thus far, we have confined our description on the records obtained on the Pacific coast. Now, the records of tide-gauges situated on the coast of Japan Sea show somewhat different aspects, when compared with those of the Pacific coast. The former is generally characterized by extremely small tides usually superposed by conspicuous secondary undulations (Pl. LXII, Fig. 3 ; Pl. LXIII, Fig. 1–4), whether the station be on the open coast or in a bay. During a cyclonic storm, the amplitude of secondary waves is generally increased and the undulation lasts for a considerable time. Comparing the records for different stations widely apart from each other, the remarkable fact is generally found that periods of conspicuous waves for different stations and for different occasions are nearly similar. It is not seldom that a series of waves recorded at one station on one occasion is quite a facsimile of that obtained in another station on another occasion. Taking the wave with the period of 20^m, which is very prevalent, its wave length corresponding to the mean depth of the sea, is calculated to be about 160 km., which is not very small, if compared with the dimensions of Japan Sea. The distance from Korea Strait to the northern end of Hokkaidô is only ten times this wave length. Hence if a train of waves be generated in or propagated into any part of this confined basin, it may be propagated impartially along the entire coast without considerable dissipation and affect all the tide-gauges on the coast. Such a train of waves may also undergo repeated reflections and refractions, before it is quite dissipated away, and cause the variety of records for different stations.

The similarity of the periods of waves accompanying cyclonic storms may, in some measure, be observed on the Pacific coast also. Besides, at different stations, though characterized by the undulations peculiar to their own, the mareograms present not unfrequently a common feature with respect to the periods of exciting waves; a fact which suggests the similarity of waves generated by and propagated from cyclonic regions. Another interesting fact is that the wave length of the most prevalent cyclonic waves is generally comparable with the dimensions of the area of the cyclonic depression. Though we are not yet enabled to solve the problem on the exact basis of hydrodynamics, it results from all probabilities that a barometric fluctuation at a cyclonic center, acting in an impulsive manner, may give rise to a train of long waves propagated from thence, and that their wave lengths are comparable with the variations at the center. N. Denison has sought to attribute the secondary undulation to Helmholtz's "Luftwogen," but the periodic variation of surface pressure is not the only means of producing a train of waves. A single impulse often suffices to generate a long train of regular waves.

Before concluding this section, a brief chronological sketch* of remarkable sea waves in Japan, recorded in connection with earthquakes, will be given below:—

1. Nov. 29, 684 A.D. A vast area of land in the province of Tosa was lost under the sea; sea waves resulted.

2. Nov. 27, 850. Sea waves on Japan Sea coast of northern Japan.

3. July 13, 869. Great waves on the coast of Sanriku.

4. Aug. 26, 887. Sea waves on the coast of Settsu and neighbouring provinces.

5. May 22, 1240. Sea waves on the coast of Kamakura in the province of Sagami.

6. Aug. 3, 1361. Sea waves on the coast of Settsu and Awa; great loss of life and property. It is recorded that on the coast of Settsu, the sea has receded for about an hour before the coming in of the wave crest, and also that the narrow strait of Naruto was drained dry.

7. Jan. 21, 1408. Sea wave recorded, but doubtful.

8. Sept. 20, 1498. Sea waves on the coasts of Kii, Ise, Mikawa, Suruga, Izu, Sagami; great loss of life.

9. Sept. 21, 1510. Sea waves on the coast of Settsu.

10. Oct. 10, 1510. Sea waves on the coast of Tôtômi. (Earthquake not recorded).

11. July 25, 1562. Sea waves on the coast of Yatsushiro, Higo; three waves counted; (no recorded of earthquake).

12. Jan. 18, 1586. Great sea waves on the coast of central part of Honshiu.

13. Sept. 1, 1596. Sea waves on the coast of Bungo; three waves counted; a few square miles of land fell beneath the sea level.

14. Jan. 31, 1605. Sea waves on the coast of Satsuma, Tosa, Kii, Ise, Tôtômi, Izu, Sagami, Awa, Kazusa, Shimôsa, Hachijô-jima. Drain-

---

*) The chief material were taken from the Report of the Earthquake Investigation Committee, No. 46.

ing of the sea before the arrival of high water is recorded on the coast of Awa, Kazusa, Shimôsa, Tòtômi, Ise; in the last province the sea had receded for about two hours before the arrival of the wave crest. Coasts of Seto-naikai and the Bay of Ôsaka were safe.

15. Dec. 2, 1611. Sea waves on the coast of Sanriku and Hokkaidô.

16. Nov. 26, 1614. Sea waves on the coast of Echigo.

17. March 1, 1633. Sea waves on the coast of Atami in Izu.

18. Oct. 30, 1662. Sea waves on the coast of Hiuga and Ôsumi; several square miles of land lost under sea.

19. April 13, 1677. Sea waves on the southern coast of Rikuchiu; at Miyako, three large waves counted in the interval ca. 10^h p.m.-2^h a.m.

20. Dec. 30, 1703. Sea waves on the coast of Sagami, Awa, and Kazusa; four waves counted.

21. Oct. 28, 1707. Remarkable great sea waves devastated the Pacific coasts of Kiushiu, Shikoku, and also of Honshiu up to Izu to the north. In Susaki, Tosa, eleven or twelve waves were counted in the interval about 3^h p.m.-4^h a.m.; at Kôchi, the wave reached so far into the land as several km.; at the same district, the third wave was the highest. Coasts of the Bay of Ôsaka were also damaged by waves.

22. Aug. 29, 1769. Sea wave in Satsuma.

23. Feb. 10, 1792. Sea wave at Shimabara in Hizen, accompanying the eruption of Mt. Unsen*; it is recorded that the high tide occurred seven or eight times a day. On this occasion a portion of the mountain was split up and fell into the sea forming a number of small isles. Inundated area, however, was confined to the coast of the Sea of Tsukushi.

24. June 29, 1799. Sea wave at Miyakoshiura, Kaga.

25. July 10, 1804. Sea waves at Kisakata; a portion of the sea upheaved into land.

26. Dec. 12, 1833. Sea wave in Sado: draining of the sea was observed before the arrival of the high water.

---

*) F. Omori, Note on the eruption of the Unsen-daké in the 4th year of Kansei (1792) Proc. Tôkyô, Math.-Phys. Soc. 4, p 32, 1907.

27. Feb. 9, 1834. Sea wave on the coast of Ishikari; three waves counted.

28. 1835. Sea wave at Hanasaki.

29. April 25, 1843. Sea waves on the coasts of Oshima, Kushiro and Nemuro; two large waves counted in 6^h a.m.-10^h a.m.

30. Dec. 23 and 24, 1854 (Ansei). Notorious sea waves on the Pacific coast.

31. Aug. 23, 1856. Sea waves occurred on the coast of Oshima and Iburi in Hokkaidô.

32. June 4, 1893. Sea waves in Kurile Islands (Chishima); at Shibetori periods of 5 waves estimated were 20^m–30^m.

33. March 22, 1894. Sea waves on the coast of Nemuro; periods of more than ten waves were estimated to have been 20^m–30^m.

34. June 15, 1896. Great sea waves on the entire coast of Sanriku.

35. Aug. 5, 1897. Sea waves on the coasts of Sanriku.

§ 9. OSCILLATION OF LARGE BAYS AND ANOMALY OF TIDES.

Thus far, we have generally treated of secondary undulations, the periods of which are much shorter than those of the principal tidal components, viz. the diurnal and semidiurnal. We will now proceed to consider those undulations of much longer periods, which commonly exist in the tides near the end of long inlets or estuaries.

Exaggeration of oceanic tides which takes places in shallow seas and in estuaries, has often been explained merely by the law of amplitude given by Green.* Airy† attempted to explain anomalies of tide observed in some rivers by the consideration that for a wave of finite amplitude different parts of the wave

*) Green, Camb. Trans. 6, 1837; Math. Papers, p. 225.

†) Airy, Tide and waves, Art. 198

profile travel with different velocities; but his argument has been proved unsustainable.* Again, inferior tidal components, known as compound tides or overtides, which become conspicuous only in shallow basins, have been explained by the analogy of combination tones in acoustics.† It seems to us, however, the theory alone is not sufficient to account for the facts that in some gulfs or bays the amplitudes of superior tides are often comparable with those of the proper tidal components and also that most pronounced compound tides are different for different bays or gulfs. Ferrel‡ attempted to explain some irregularities of oceanic tides by considering oceans as making stationary oscillation as in the case of seiches in lake. Recently R. A. Harris,** according to a similar view, constructed a cotidal chart of the world. In his theory, water on the globe is divided into several distinct portions, each of which has the proper period of its own stationary oscillation. This point have been subjected to criticism by G. H. Darwin. He applied his theory also to the explanation of tidal phenomena in many bays and straits, the standing oscillations of which are forced by tidal waves incident on their mouths. He considered, however, exclusively the forced oscillation with diurnal and the semidiurnal periods, and has not considered those oscillations peculiar to each bay.

Now according to our view, any bay or gulf, either small or large, may be considered as a resonator which oscillates with its own period, if it be excited by waves in the external sea having the synchronizing component. If the proper period of the bay happens to coincide nearly with one of the tidal

---

*) McCowen, Phil. Mag. (5), **35**, 1892.

†) W. Thomson, Proc. Roy. Soc. **7**. G. H. Darwin, Brit. Ass. Rep. **188**.

‡) Ferrel, Tidal Researches, Rep. Coast and Geodetic Survey, Washington, 1874.

**) R. A. Harris, Manual of Tides.

components, that component will become more or less prominent, according to the degree of proximity of the proper and the exciting period. Again, it is possible that astronomical components of superior orders, compound tides resulting from the combination of several astronomical components or many indefinite components arising from meteorological causes, may sometimes become prominent by the resonance of bay water, though almost insensibly small in open sea. Moreover, the case may occur, in which a solitary wave of wide extent caused by some disturbances either meteorological or geotectonic, excites the oscillation of a long period proper to a bay. These oscillations in a bay will more or less deform the tidal curve and cause an anomaly of tides peculiar to that bay. According to this view, the proper periods as given by our formula were calculated for different bays or gulfs, which are notorious for abnormal range of tide, and also those for which some remarkable irregularities of tide was observed in the mareograms given in the Report of the Krakatoa eruption often cited.

The results of calculations, together with their bearings upon the tidal irregularities for a number of bays, will be given below.

(*a*) Bay of Fundy, Canada. Near the end of this bay, spring tides range 15 m., while near its entrance the rise is only 2.5 m. to 3.5 m. For the calculation, the mouth line was taken from Cape Cod to Cape Sable; and the end of the Bay was taken at Port Greville. Then $l=460$ km., $h=141$ m.; hence $T=13.0^{h}$. If the mouth line be taken between Yarmouth and Machias, the calculated period is 11.6^{h}. In any case, the proper period of this bay would be very near 12 hours. The abnormal high tide may then partly be explained by the coincidence of the proper period with one of the semidiurnal tides, though the tidal phase is slightly retarded toward the end of the Bay and therefore the phenomenon can not be wholly attributed to the standing oscillation.

(*b*) Bay of Bengal. Near the mouth of this bay, the tidal range is small, being less than half a meter at the southern coast of Ceylon, while in the bay, the range is 1–2.7 m., increasing abruptly near the end of the Bay. Since the tidal phase is nearly the same for the great part of the bay, principal part of the tide in the Bay is probably due to the standing oscillation of it. Taking the mouth line from the eastern coast of Ceylon to the northern end of Sumatra, and the end of the Bay at Akyab in Burma, we obtain $l=1500$ km. and $h=1950$ m., which gives $T=12.0^h$, coinciding with the period of semidiurnal solar tide.

(*c*) Madura Strait, Java. Mareograms of Ujong Sourabaya and Karang Kleta reproduced in the Krakatoa Report of the British Association show very marked irregularities of tide in the narrow strait of Sourabaya, which connects the end of wide Strait of Madura with the Java Sea. Since the width of the Strait of Sourabaya is extremely small in comparison with that of the Madura Strait, we may consider the latter strait as a rectangular bay, having its end inside the former strait. The effect of narrow opening at the end of a bay is only to slightly reduce the effective length of the bay, or in other words to shorten the period of oscillation by a small amount. The length of the Madura Strait, considered as a bay is about 160 km. As for the mean depth of the basin, no chart of a sufficient scale is at hand, so that we may only roughly estimate it from the data collected from different maps* available. In any case,

*) Maps of Encyclopedia Britannica, Berghaus' Physikal. Atlas, etc.

however, the mean depth can not differ much from 30 m. The period calculated from these data is 8.8^h. In order to see, if any long wave corresponding to this result of calculation exists actually in record, the mareogram of Karang Kleta was analysed by means of the tide-rectifier. Eliminating the great parts of diurnal and semidiurnal components, the remaining curve reveals to us an evident trace of waves with the mean period of 8.0^h (Pl. LXIV, Fig. 1).

The exciting cause of this wave is probably the compound tide usually denoted by MK, the period of which is about 8.2^h, nearly coinciding with the proper period of the Madura Strait.

(*d*) Port Adelaide, Australia. A mareogram (Pl. LII, Fig. 3) of this Port given in the Krakatoa Report shows remarkable change of diurnal inequality on successive days. On eliminating the principal parts of diurnal tides, the resulting curve shows apparent beats of semidiurnal tides (Pl. LXIV, Fig 2). The period of the characteristic component which forms beats with the usual semidiurnal tides, is about 10.9^h, as estimated from the rectified curve. It is probably to be attributed to the standing oscillation of the St. Vincent Gulf. Taking the mouth line from Trowbridge Point to Cape Jevis, and the end of the Gulf at Wakefield, we have $l=140$ km. The estimated mean depth is 21.5 m. Hence $T=10.8^h$, which coincides very well with the observed period.

(*e*) Port Phillip, Australia. Mareogram (Pl. LII, Fig. 4) of Williamstown given in the above quoted Report, shows some irregularities of tide which suggests the existence of very long undulation peculiar to the bay. The curve was rectified (Pl. LXIV, Fig. 3) and a period of 8.3^h, was detected. Now the period of seiches in this nearly enclosed basin can by no means become so long as 8^h, unless the mean depth of the Port be less than one meter. This period is probably due to the undulation of the whole basin with its narrow neck communicating with open sea. For the calculation of the period corresponding to this mode, the necessary data were estimated from the chart given in Harris' Report: $S=13.7\times10^8$ m²., $l=2.95$ km., $b=4.08$ km., $h=18.4$ m. and $\lambda=200$ km. The period calculated from these data is 8.39^h which almost coincides with the observed period.

Beside the above enumerated examples, there are many bays or straits, the forms of which seem favourable for their own standing oscillation of long periods, and in which the ranges of tides are comparatively large. The periods of these oscillations were estimated as follows:—

Bay or Strait.	T in Hours
Adriatic Sea.	15
Mozambique Channel.	7
Bristol Bay, Alaska.	15 and 28
Hecate Strait, British Columbia.	7
Hudson Strait, Canada.	11
Bristol Channel, England.	5
Gulf of St. Malo, France.	7

In conclusion, we wish to express our best thanks to Prof. H. Nagaoka under whose supervision the entire work has been carried out and who gave us much useful instruction throughout the course of our investigations. Equal thanks are due to Prof.

F. Omori who favoured us with valuable instruction and with the records of numerous earthquake sea waves. We also wish to express our thanks to Assist. Prof. A. Imamura, and to Mr. M. Sugiyama of the Military Staff, who kindly placed valuable mareograms at our disposal. Lastly, our cordial thanks are due to Messrs. S. Iwamoto, N. Watanabe, T. Hirata, Y. Inouye and T. Fukuda who have been zealous cooperators in the course of our observations.

NAME LIST OF STATIONS.

A

Aburatsu 油津
Aburatsubo 油壺
Adashika 新鹿
Akashiseto 明石瀬戸
Akkeshi 厚岸
Amai 雨井
Aodashi 青出
Aomori 青森
Ariake 有明
Atami 熱海
Awa (Tosa) 安和
Ayukawa 鮎川

B

Bisanseto 備讃瀬戸
Bokkiriso 暮露礎

C

Chôshi 銚子

D

E

Etchiujima 越中島

F

Fukahori 深堀
Fushiki 伏木
Futami (Sado) 二見
Futami (Ogasawarajima) 二見

G

H

Hakodate 函館
Hamanaka 濱中
Hanasaki 花咲
Heda 戸田
Heida 平田
Heshima 戸島
Hinoura ヒノ浦
Hiroshima 廣島
Hososhima 細島
Hosoura 細浦

Ibusuki 揖宿
Ikeda 池田
Imazu 今津
Inuboye 犬吠
Isegahama 伊勢ヶ濱
Isenoumi 伊勢海
Iwasaki 岩崎
Iwaya 岩屋

J

K

Kagoshima 鹿兒島
Kajiki 加治木
Kamagôri 蒲郡
Kamaishi 釜石
Kamagasaki 鎌崎
Kameura 甕浦

Kamezaki	龜崎
Kamiiso	上磯
Kanaiwa	金石
Kanegafuchi	鐘ヶ淵
Karatsu	唐津
Kashiwazaki	柏崎
Katsuura	勝浦
Kelung	基隆
Kiritapp	霧多布
Kishiwada	岸和田
Koishihama	小石濱
Kojirohama	小白濱
Kokabe	コカベ
Konpaku	根白
Kukiura	九木浦
Kure	久禮
Kushimoto	串本

M

Maisaka	舞坂
Mera-koura	妻良子浦
Mikawa wan	三河灣
Misaki	三埼
Miyako	宮古
Miyazu	宮津
Moroiso	諸磯
Mororan	室蘭

N

Nagasaki	長崎
Naishobama	內所濱
Naoetsu	直江津
Naruto	鳴門
Nemuro	根室
Niigata	新潟
Niiyama	新山
Nomi	野見
Nonomae	野々前

O

Ôdsuchi	大槌
Ôfunato	大船渡
Ôgeura	大毛浦
Ôishi	大石
Okirai	越喜來
Okuchi	奥地
Omaezaki	御前崎
Ôminato	大湊
Onshantin	元山津
Ôsaka	大阪
Ôtani	大谷
Otaru	小樽

P

Pusan	釜山

R

Ryôishi	兩石
Ryôri	綾里

S

Sagami	相摸
Sakurajima	櫻島
Samé	鮫
Sasuna	佐須奈
Senzai	千歳
Shido	志度
Shimizu (Suruga)	清水
Shimizu (Tosa)	清水

Shimoda	下田
Shimonoseki	下ノ關
Shiogama	鹽釜
Shionomisaki	潮岬
Shioyasumi	潮憩
Shiraiwa	白岩
Shishimi	鹿見
Shitama	舌間
Shizuura	靜浦
Sunagozaki	砂子崎
Suruga wan	駿河灣
Susaki	須崎

T

Tachimachizaki	立待岬
Tago	田子
Takonoura	蛸ノ浦
Takow	打狗
Tei	手結
Tateyama	館山
Toba	鳥羽
Tôkyô	東京
Tomikawa	富川
Tonoura	外ノ浦
Tsuruga	敦賀

U

Uchinoumi	內ノ海
Ujina	宇品
Umegahama	梅ケ濱
Uwajima	宇和島

Y

Yamasakibana	山崎鼻
Yawatahama	八幡濱
Yebisu	夷
Yebisujima	夷島
Yedajima	江田島
Yonozu	米水津
Yoshihama	吉濱
Yotsu	輿津
Yura	由良
Yuraseto	由良瀨戶

W

Wajima	輪島
Washinosu	鷲ノ巢
Weihaiwei	威海衛

MAREOGRAMS.

PL. I-LXIV.

EXPLANATIONS.

Horizontal scale below the diagrams gives the time, generally in hours, if not otherwise stated. In a number of cases, actual hours of day are shown, midnight and noon being inscribed. The length of one hour in records of Kelvin's tide-gauge is about twice that of our instrument.

Vertical scale gives the height of level in cm.

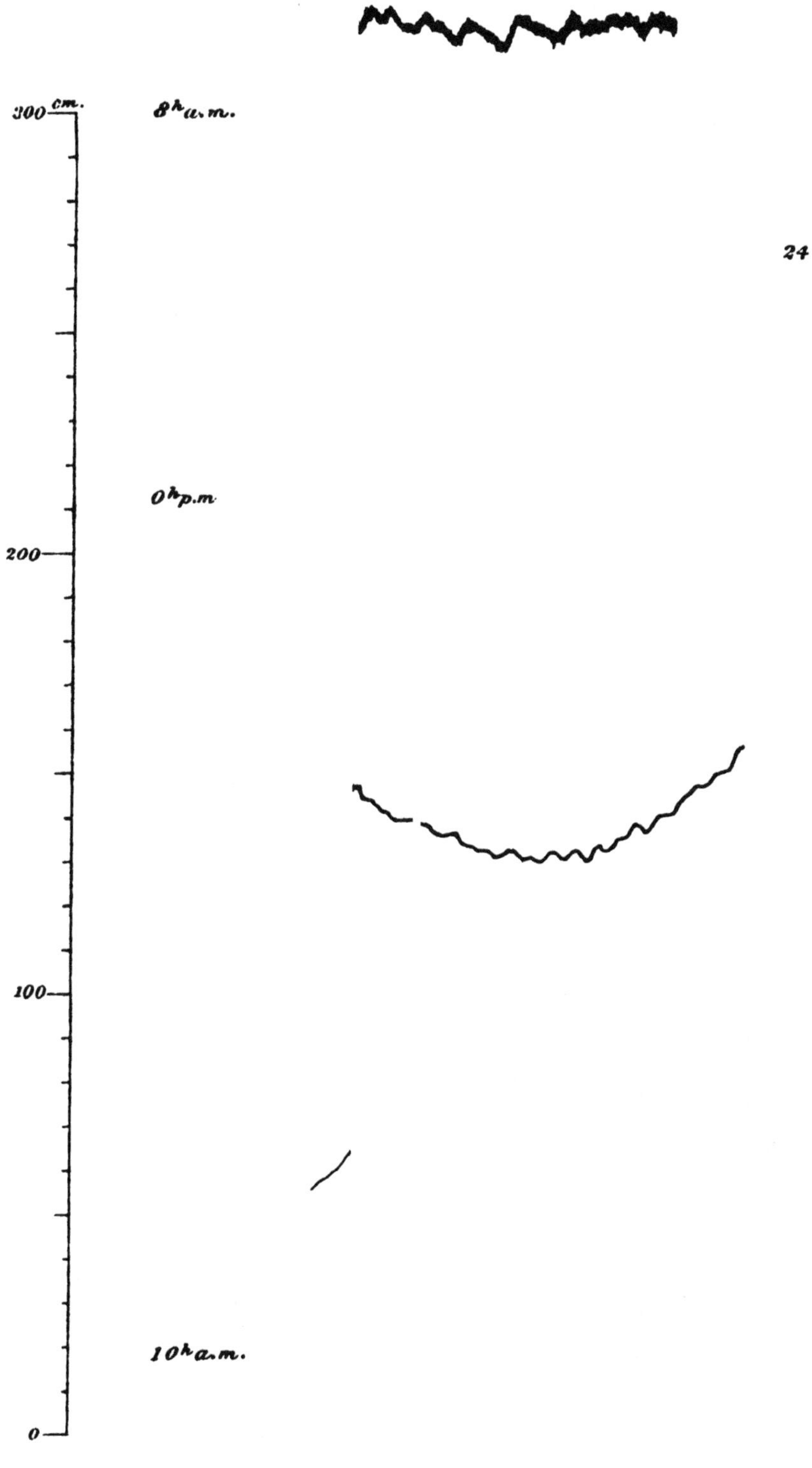

Pl. I.

300 cm. — 8h a.m.

24

0h p.m.

200

100

10h a.m.

0

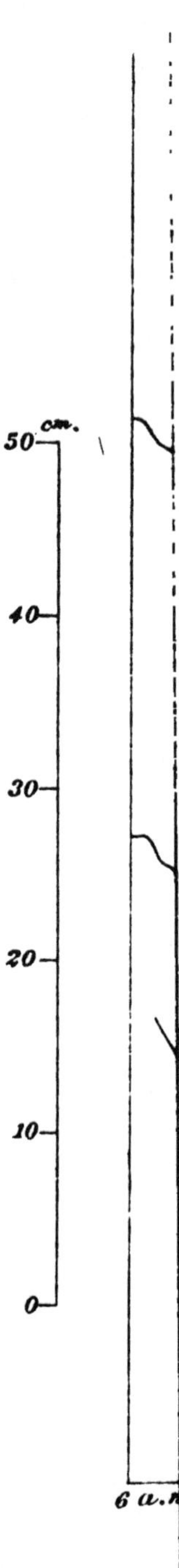
50 cm.
40
30
20
10
0
6 a.m

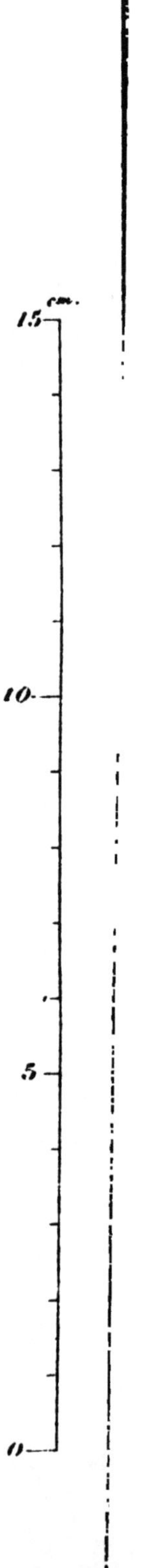
cm.
15
10
5
0

90^m (Time scale for Fig. 1.)

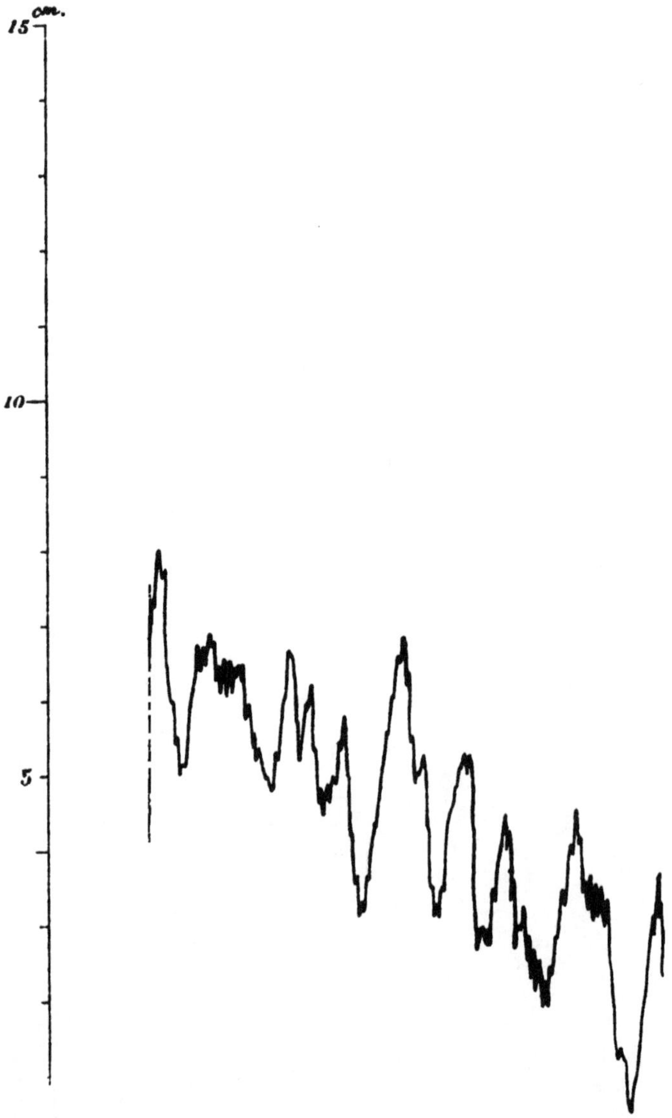

15 cm.

10

5

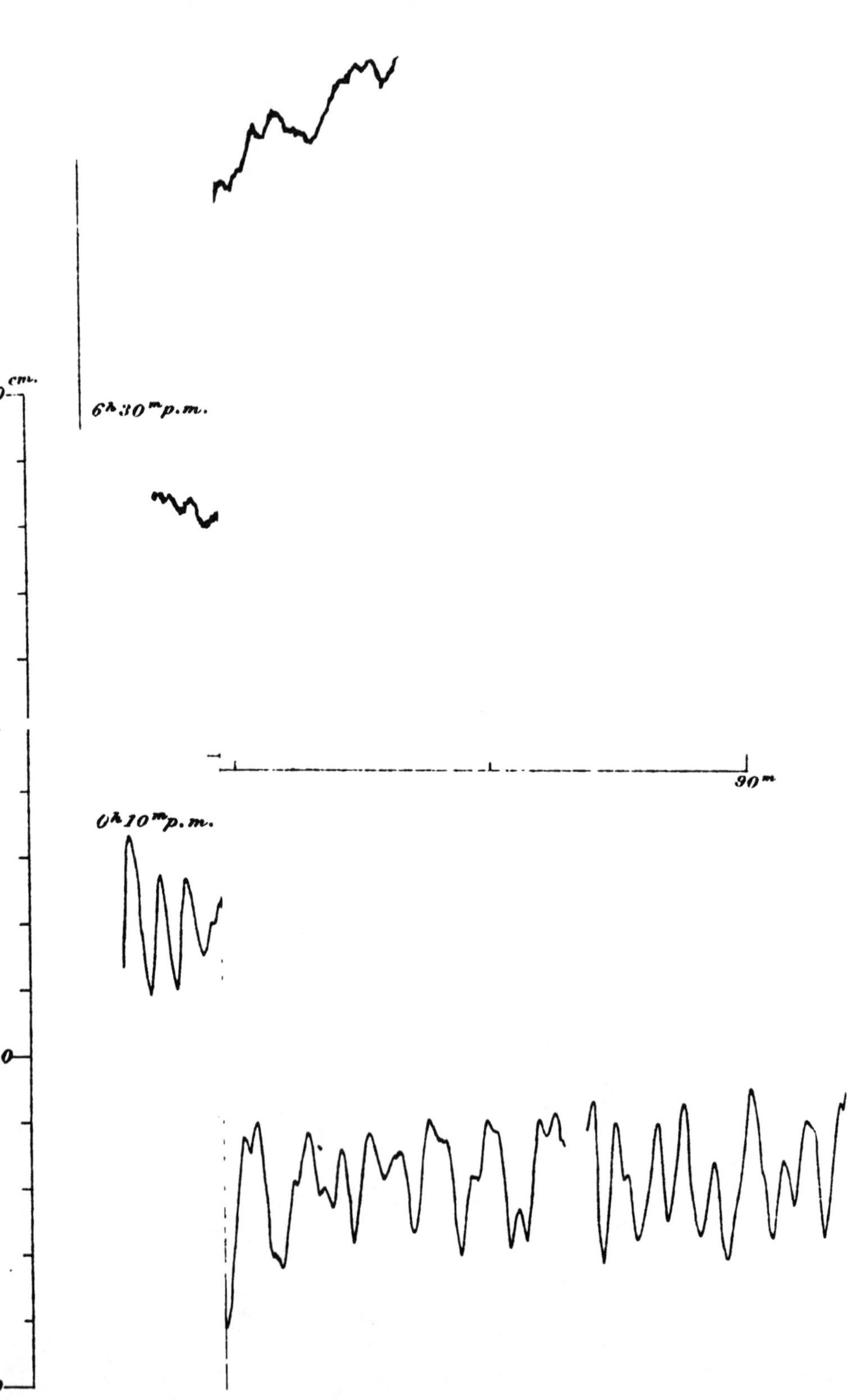
0 cm.
6h 30m p.m.
90m
0h 10m p.m.
50
0

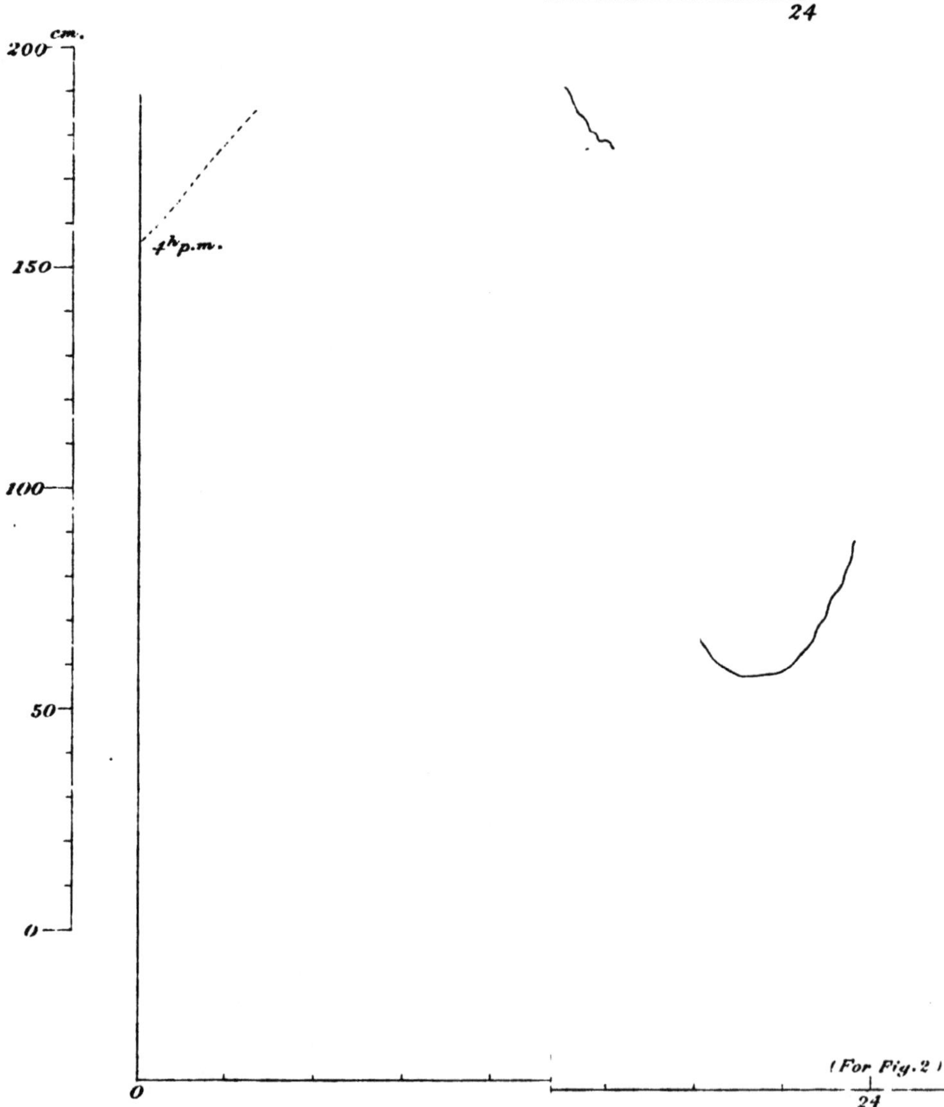
(Time scale for Fig. 1)
24
cm.
200
150
100
50
0
4h p.m.
0
(For Fig. 2)
24

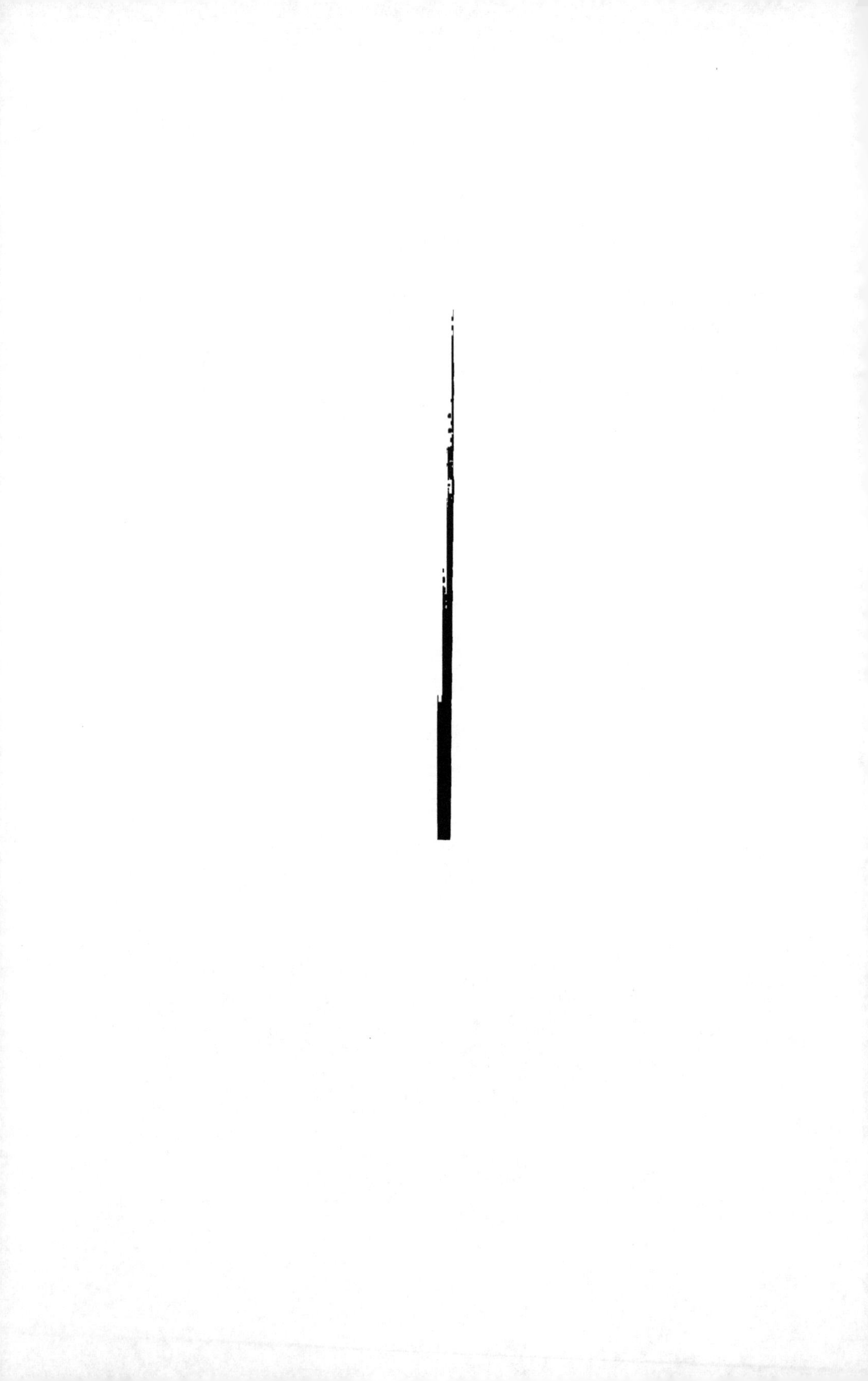

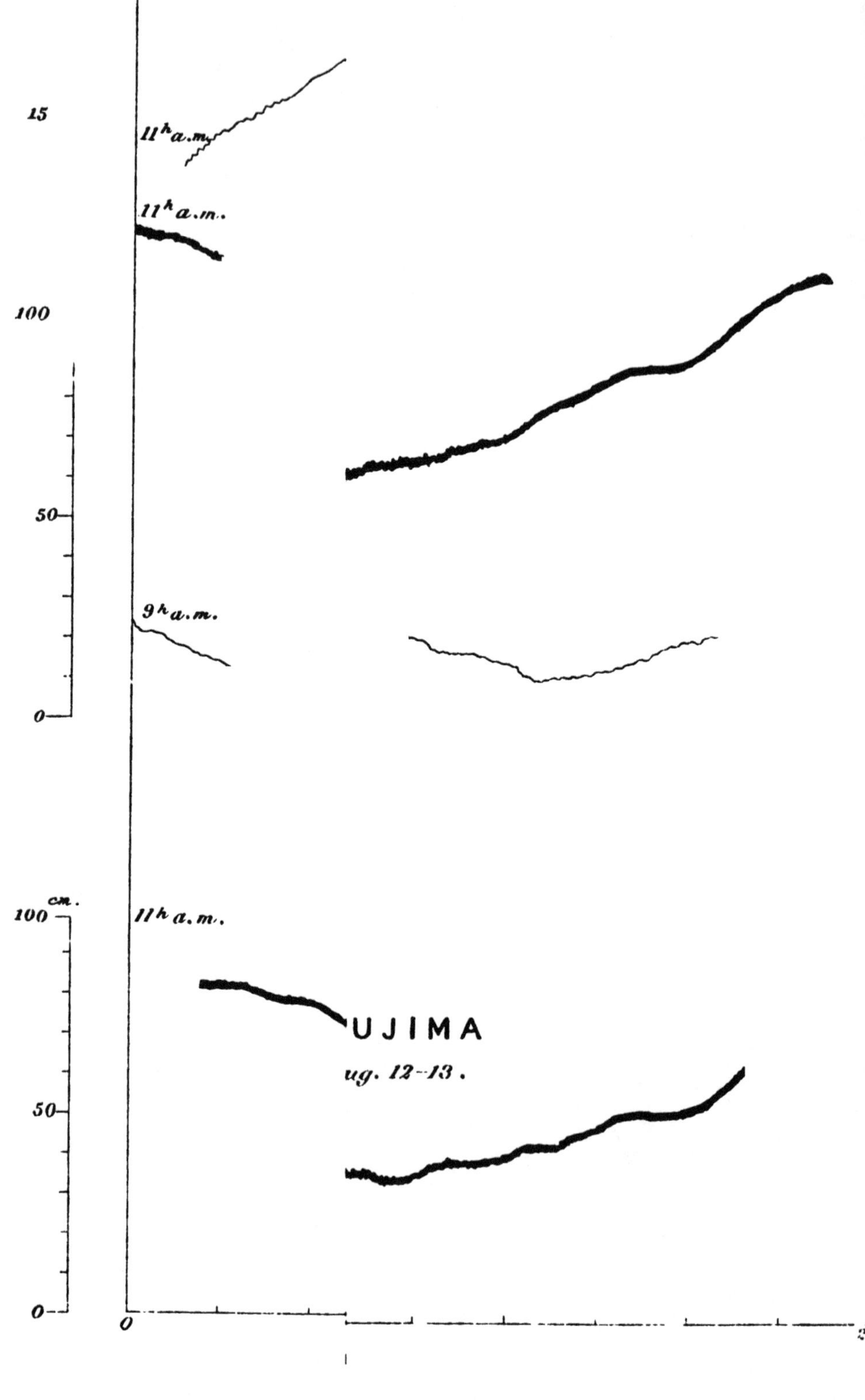

15
11h a.m.
11h a.m.
100
50
9h a.m.
0
cm.
100
11h a.m.
UJIMA
ug. 12–13.
50
0
0
24

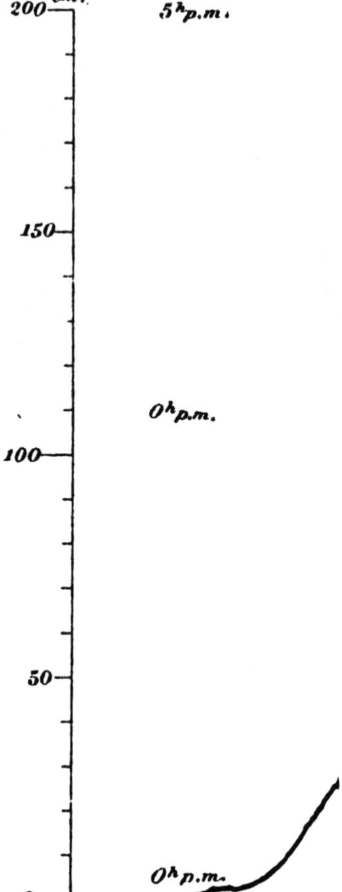

0

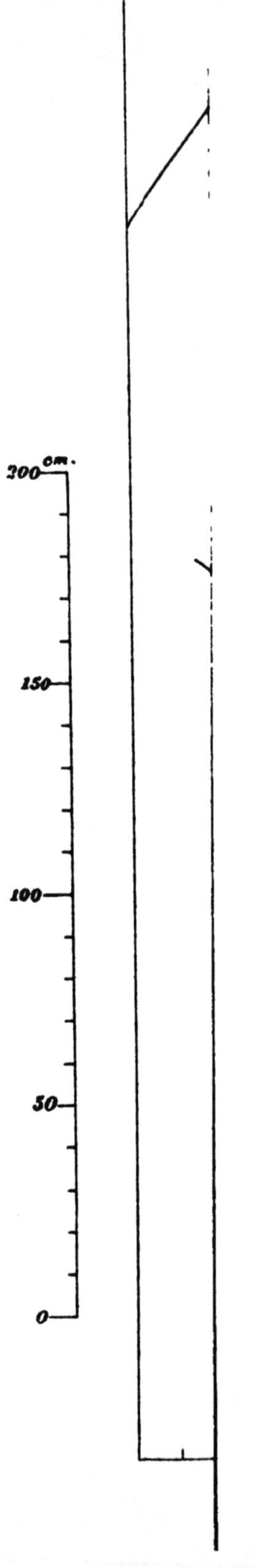

200 cm.
150
100
50
0

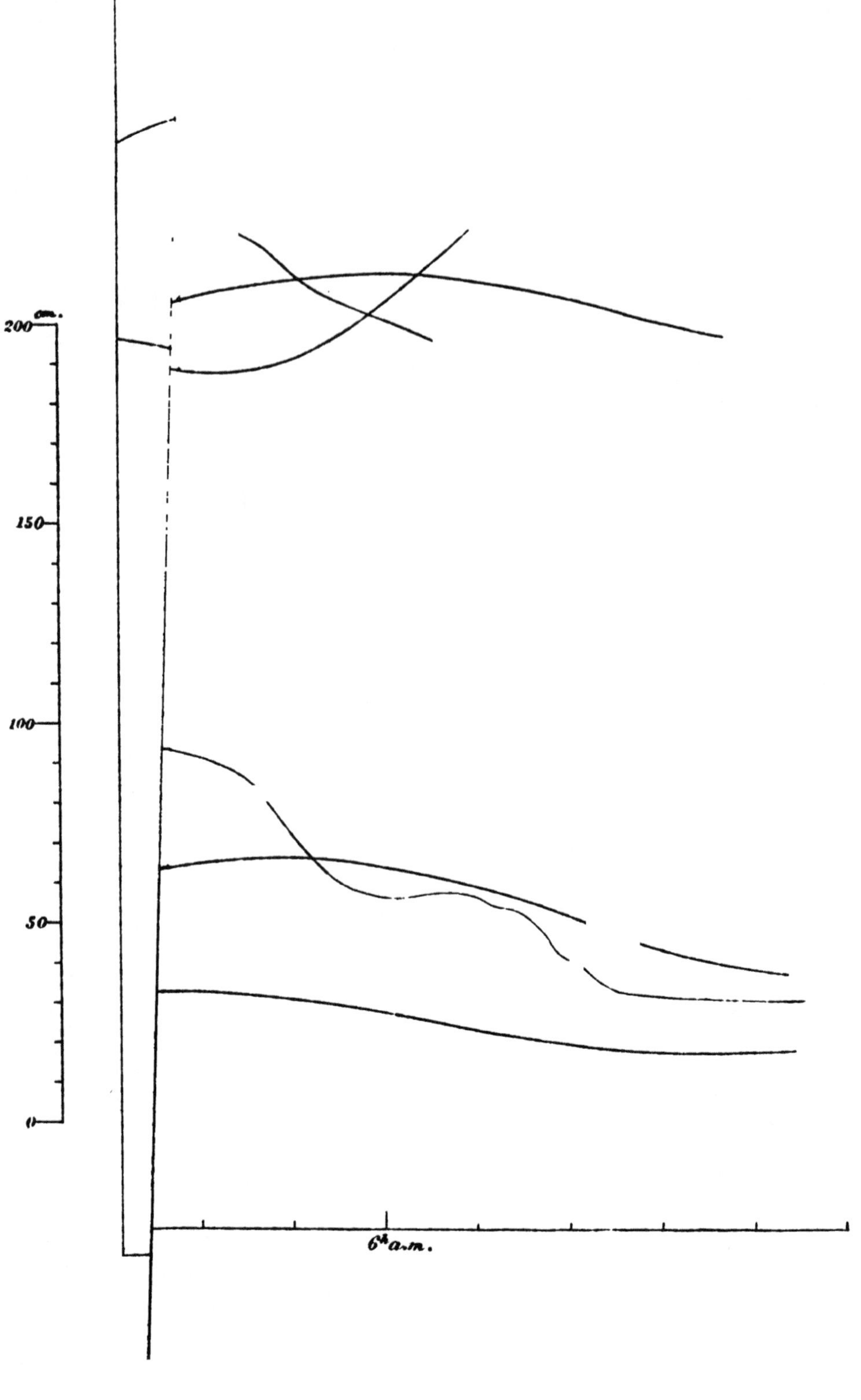

cm.
200
150
100
50
0
6h a.m.

Pl. XXXIII.

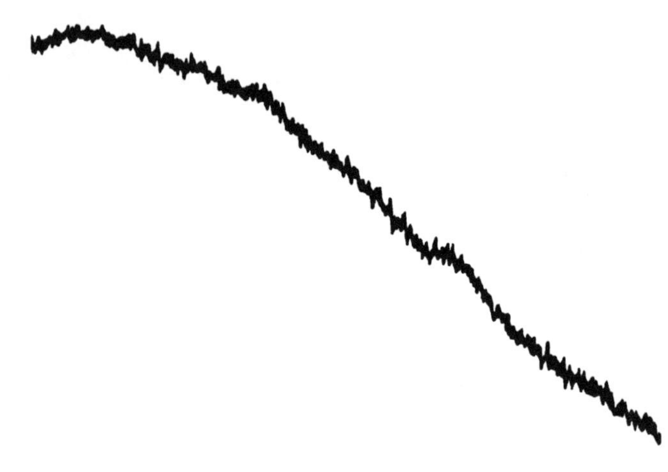

250 cm.

200

150

100

50

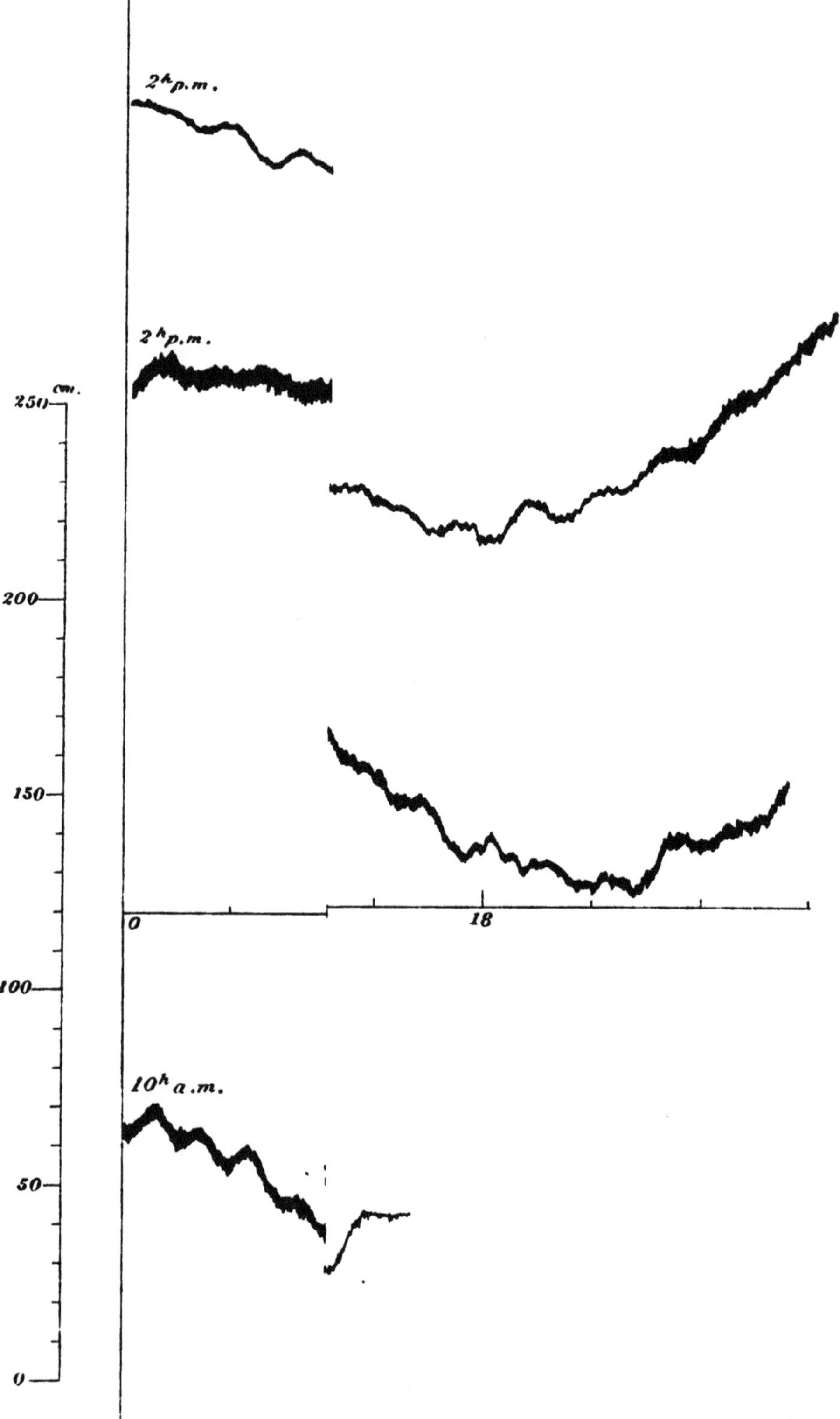
2ʰ p.m.
2ʰ p.m.
cm.
250
200
150
100
50
0
0
18
10ʰ a.m.

Aug. 17, 1903.

18

24

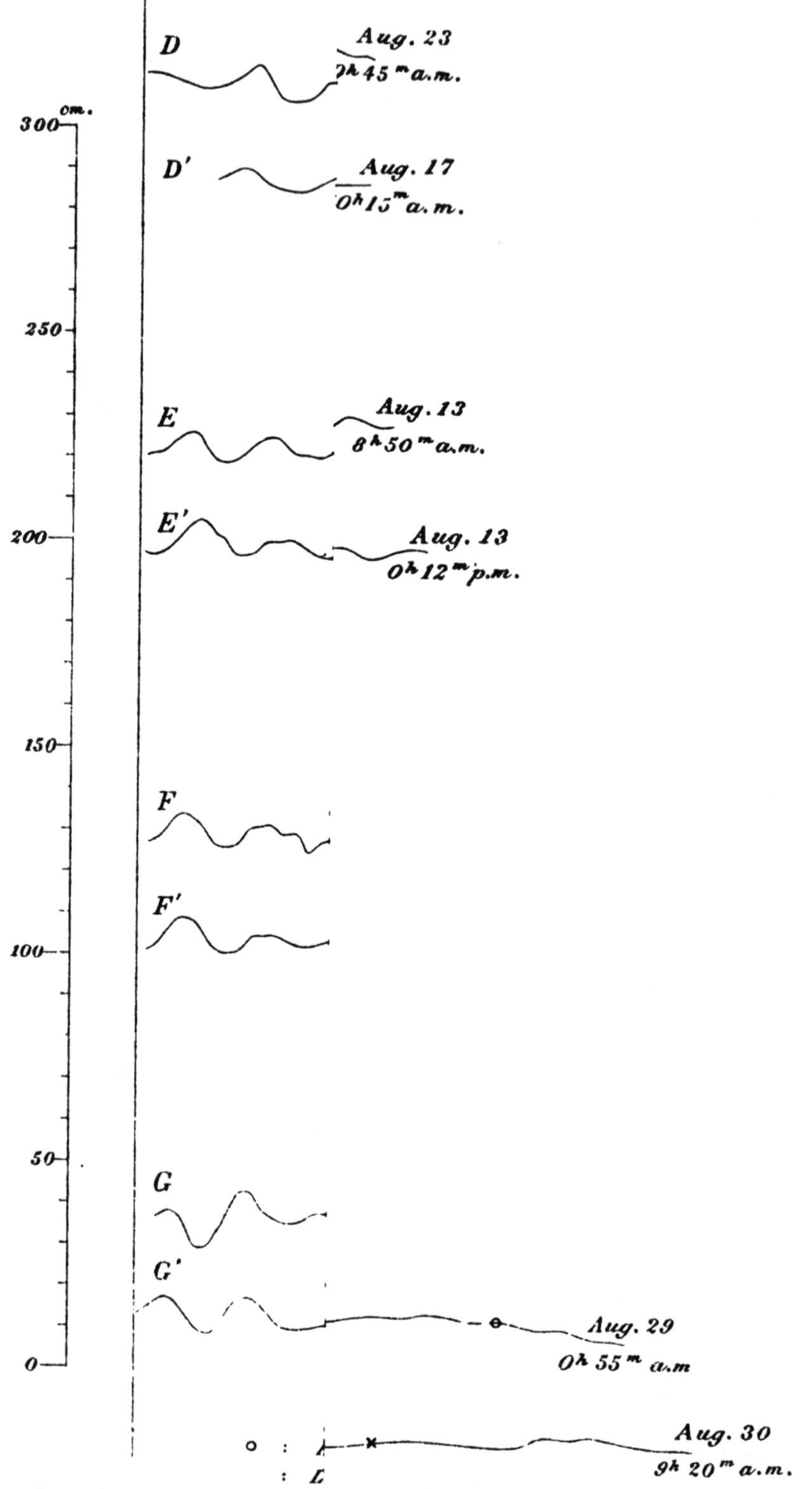
D
Aug. 23
9h 45m a.m.
300 cm.
D'
Aug. 17
0h 15m a.m.
250
E
Aug. 13
8h 50m a.m.
E'
200
Aug. 13
0h 12m p.m.
150
F
F'
100
50
G
G'
Aug. 29
0h 55m a.m
0
Aug. 30
9h 20m a.m.

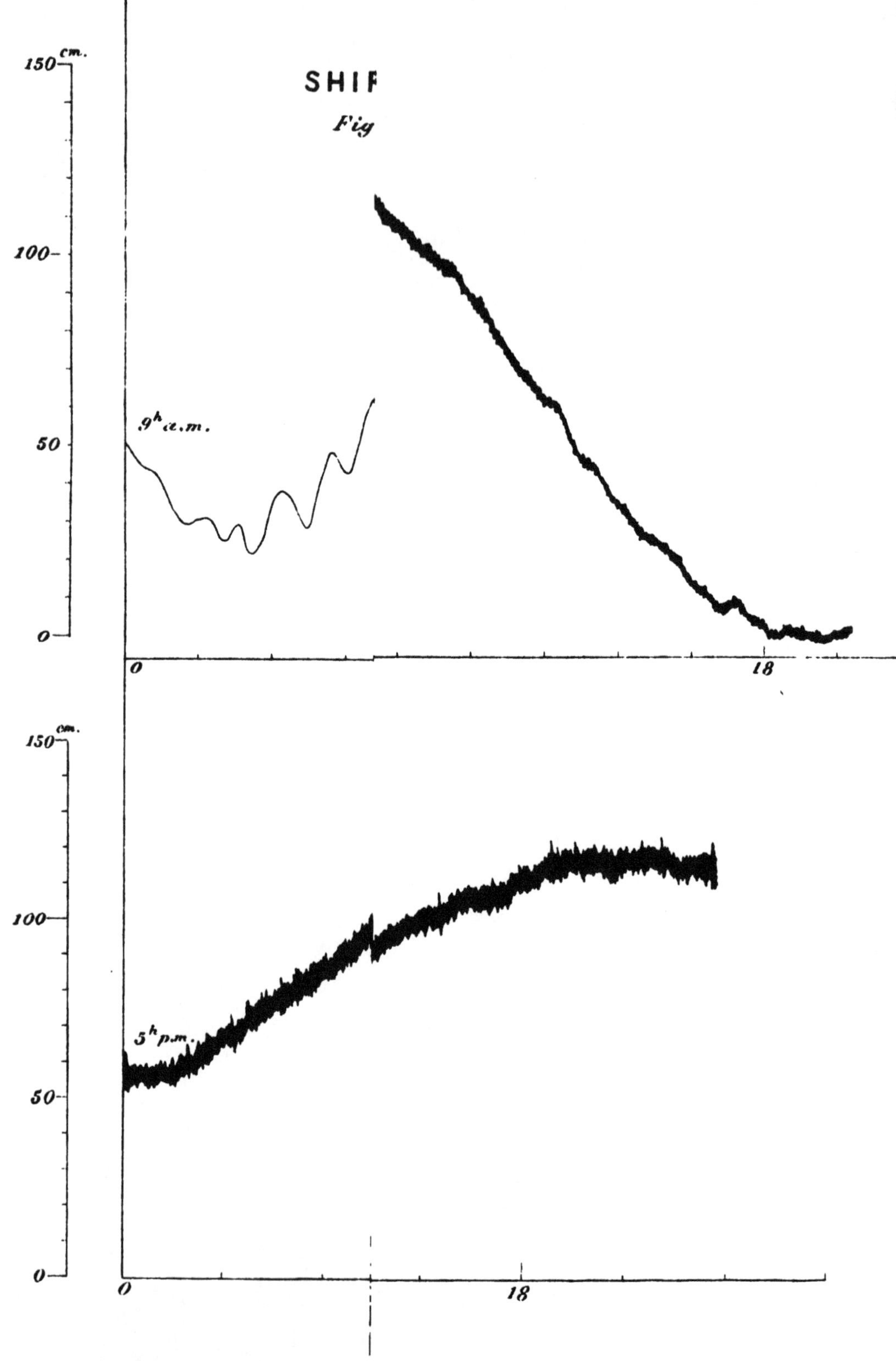

SHIF
Fig
150 cm.
100
50
0
9h a.m.
0
18
150 cm.
100
50
0
5h p.m.
0
18

cm.
250
200
150
100
50

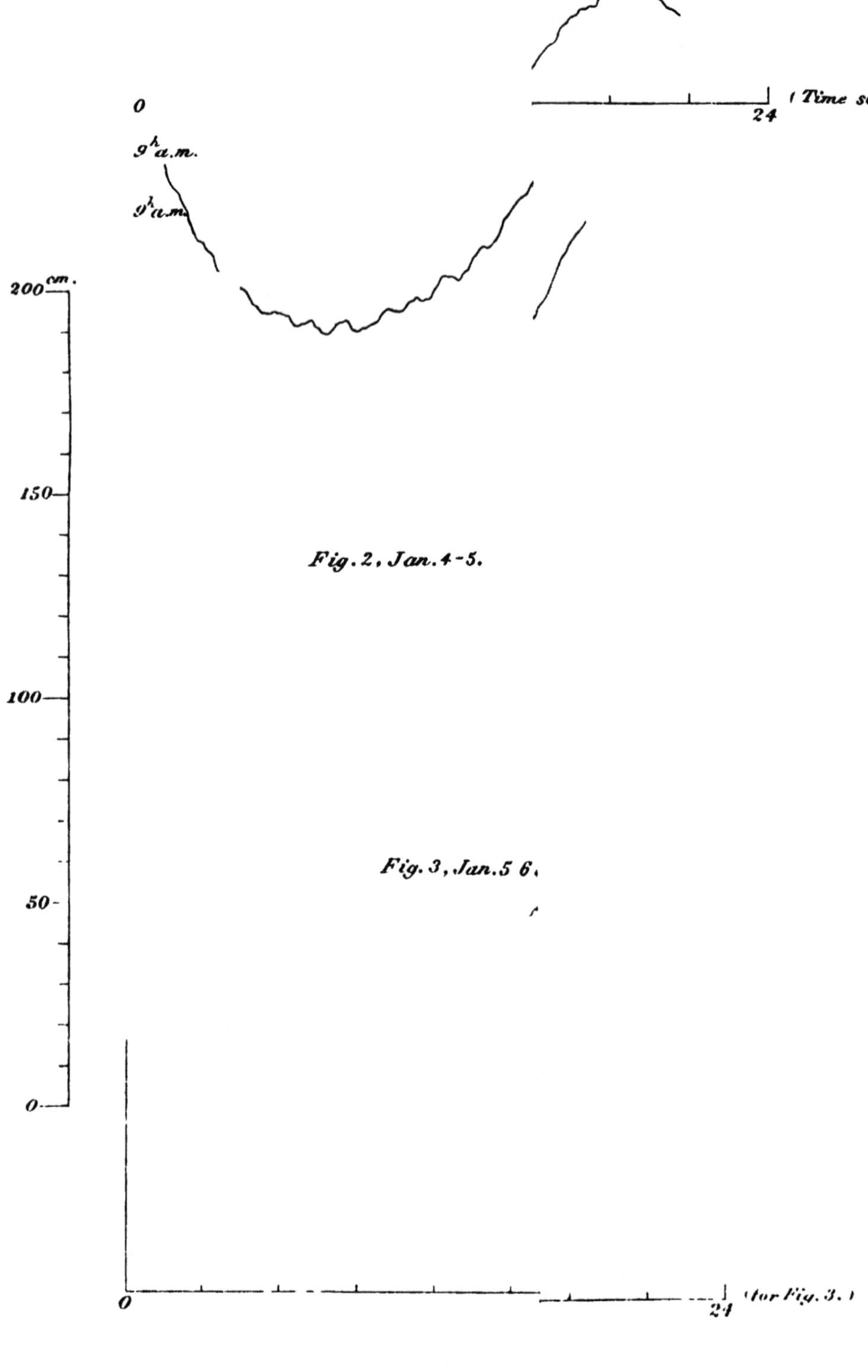

Fig. 2, Jan. 4-5.

Fig. 3, Jan. 5 6.

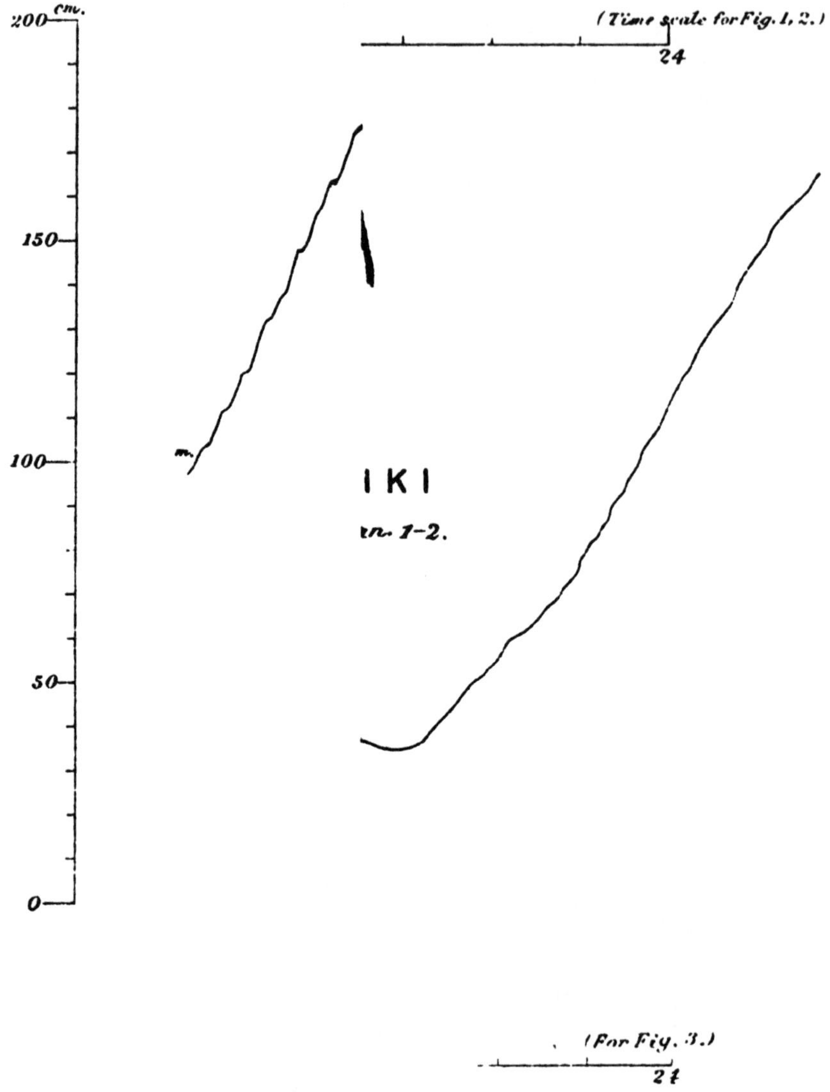

200 cm.

(Time scale for Fig. 1, 2.)

24

150

m.

100

IKI

n. 1-2.

50

0

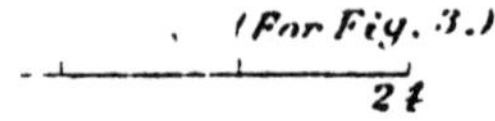

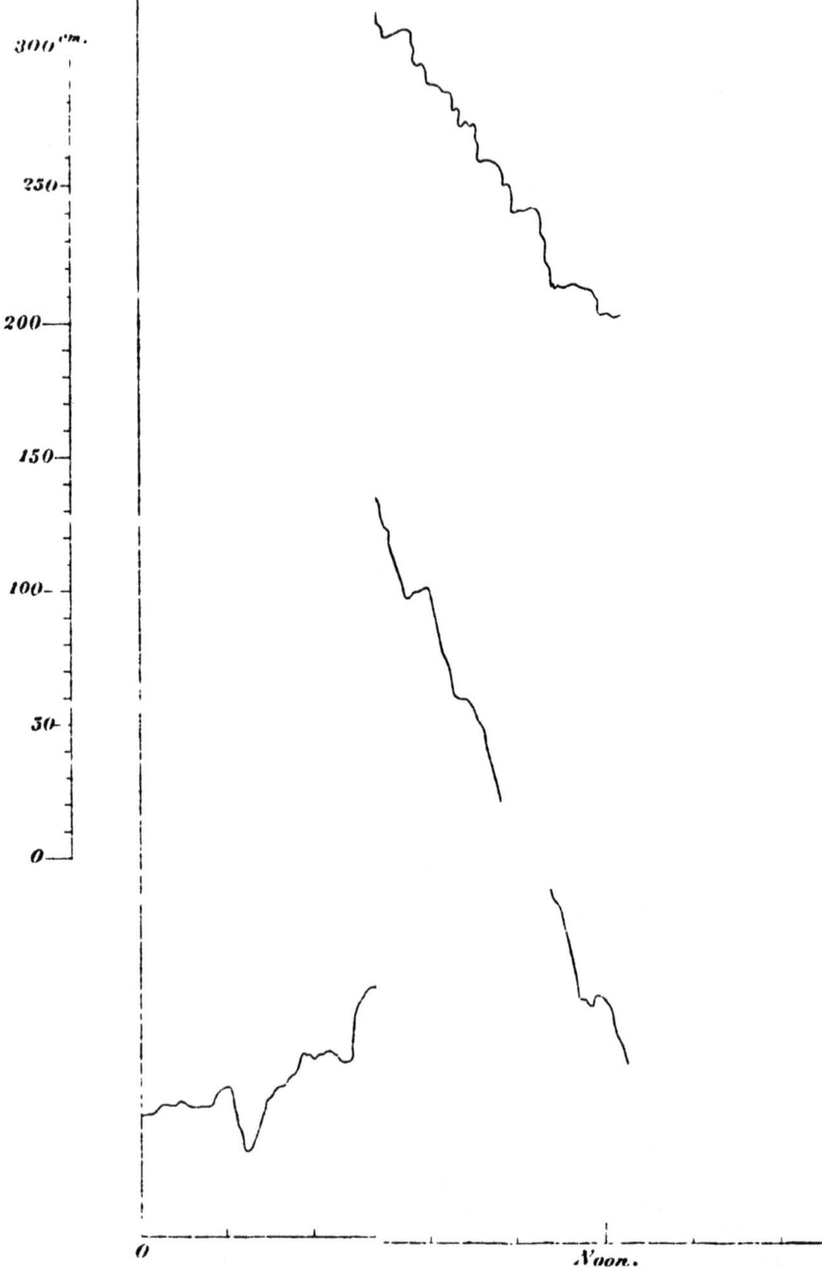
300cm.
250
200
150
100
50
0
0
Noon.

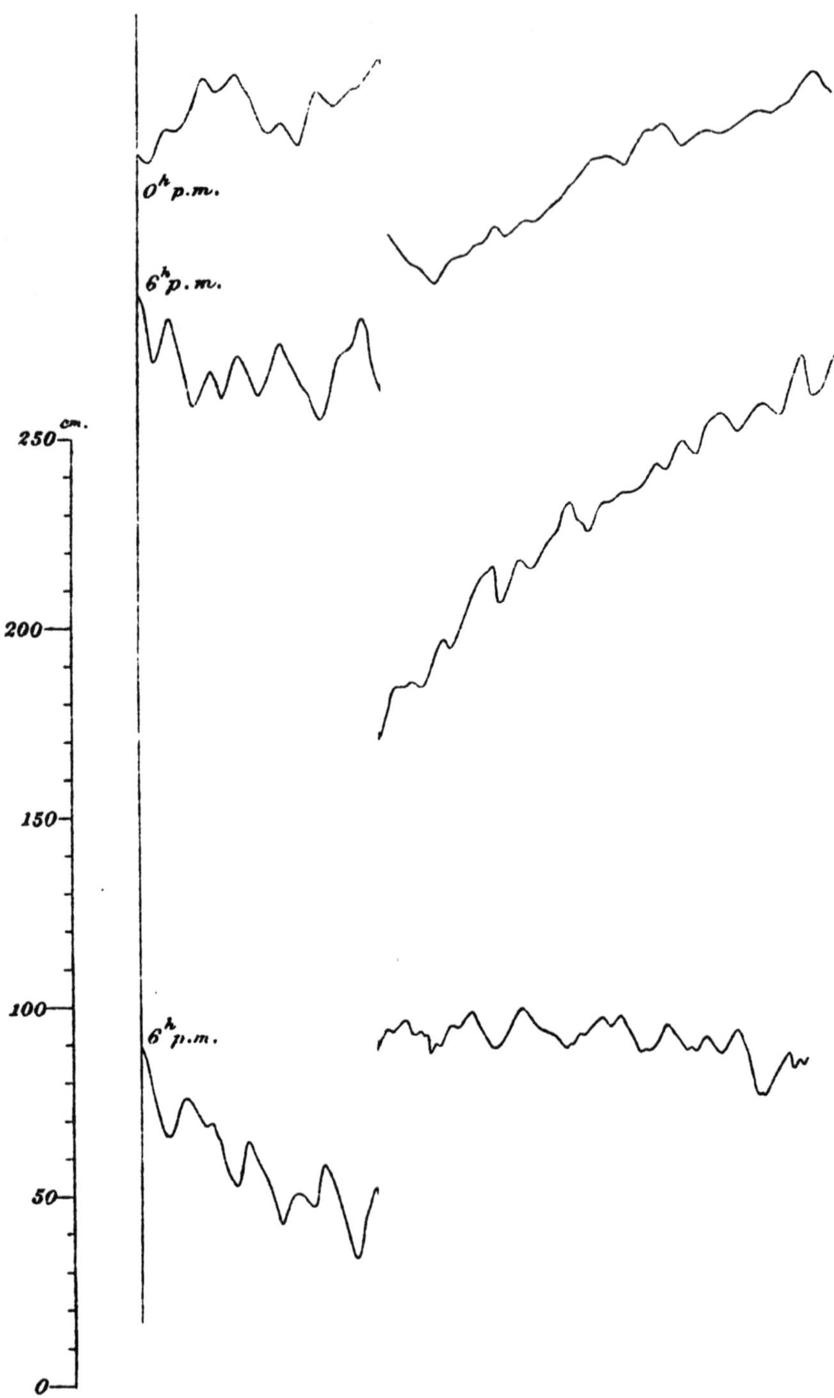
0h p.m.
6h p.m.
6h p.m.
250 cm.
200
150
100
50
0

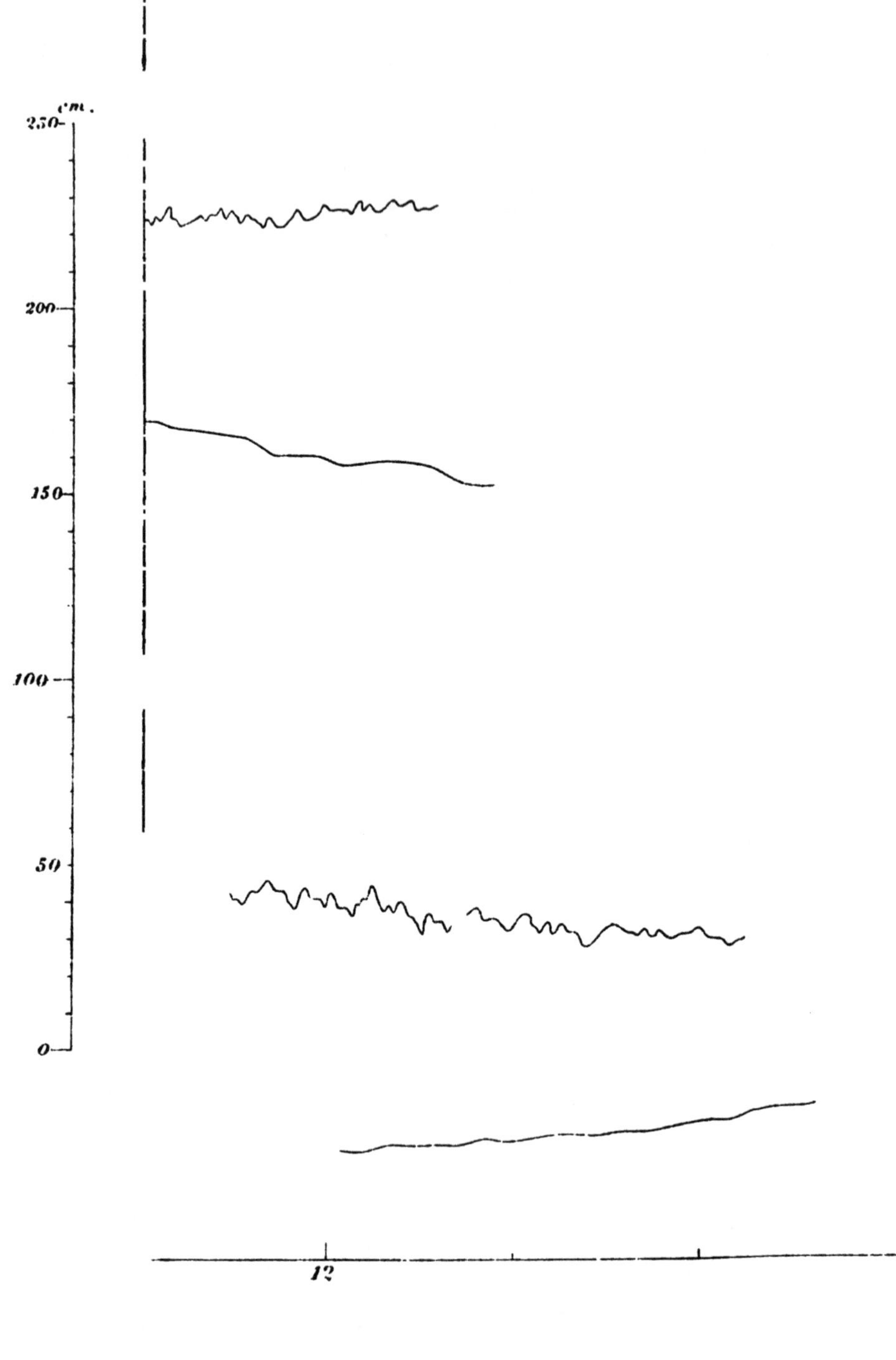
cm.
250
200
150
100
50
0
12

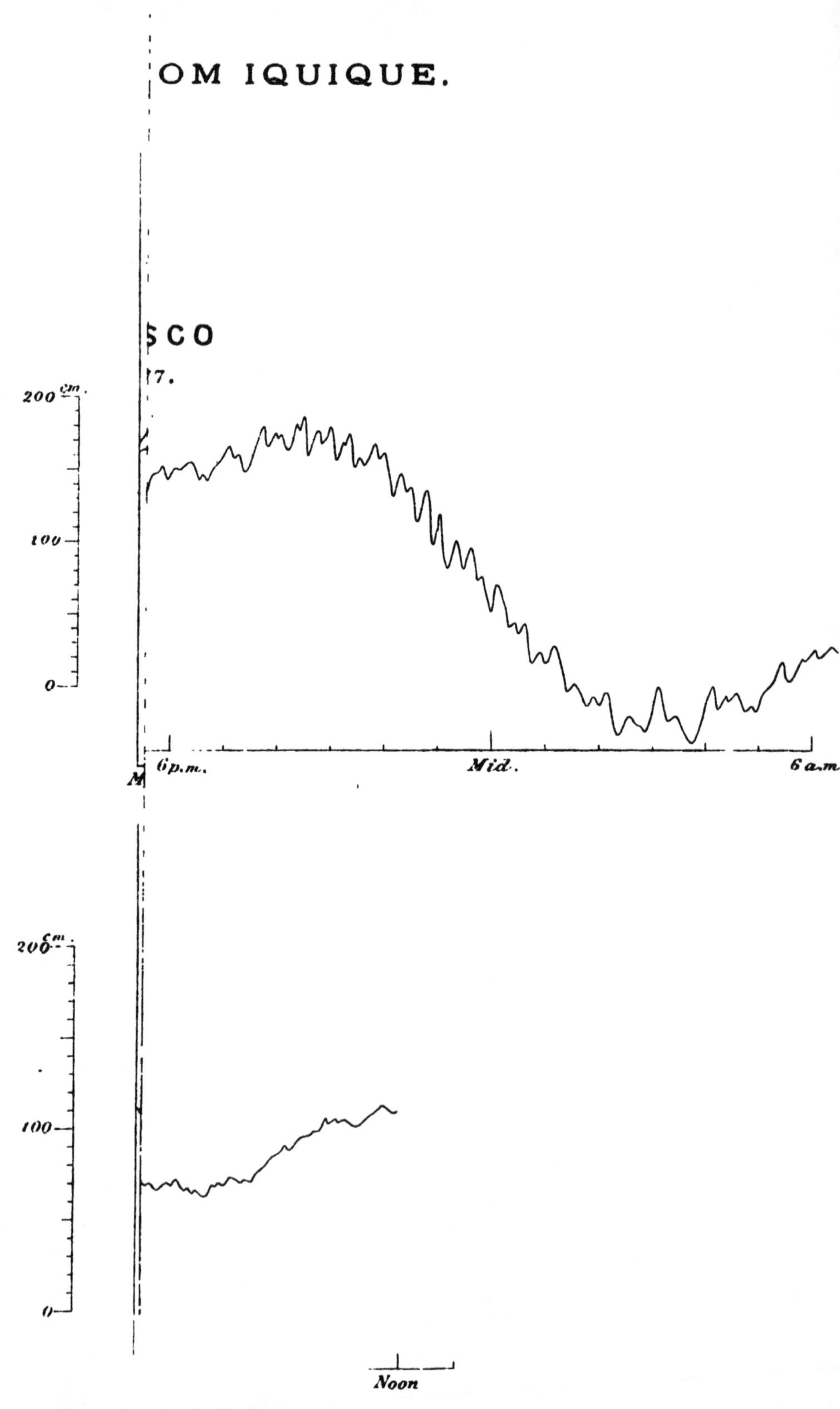
OM IQUIQUE.
SCO
77.
200 cm.
100
0
M
6p.m.
Mid.
6a.m.
200 cm.
100
0
Noon

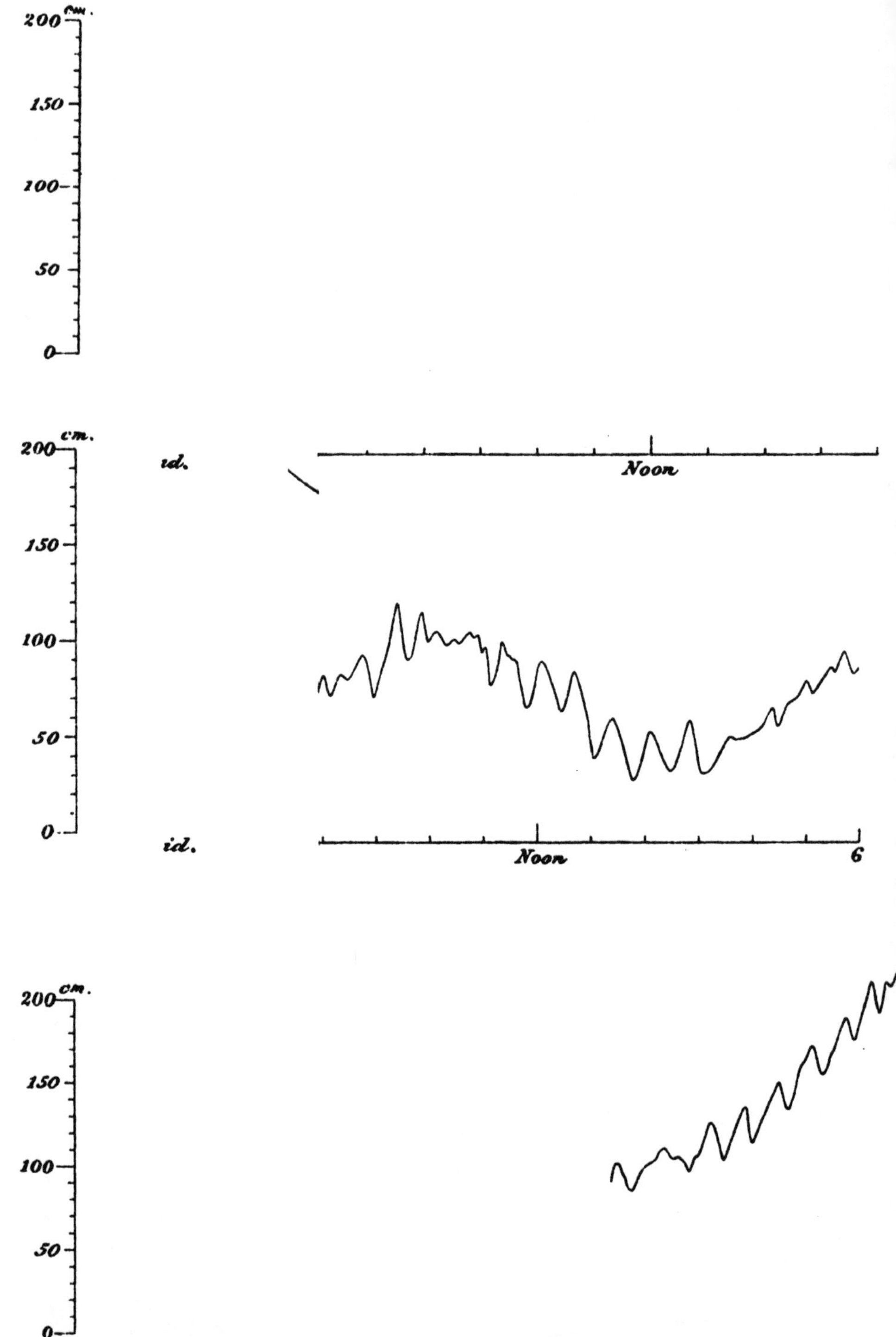

200 cm.
150
100
50
0
200 cm.
id.
Noon
150
100
50
0
id.
Noon
6
200 cm.
150
100
50
0

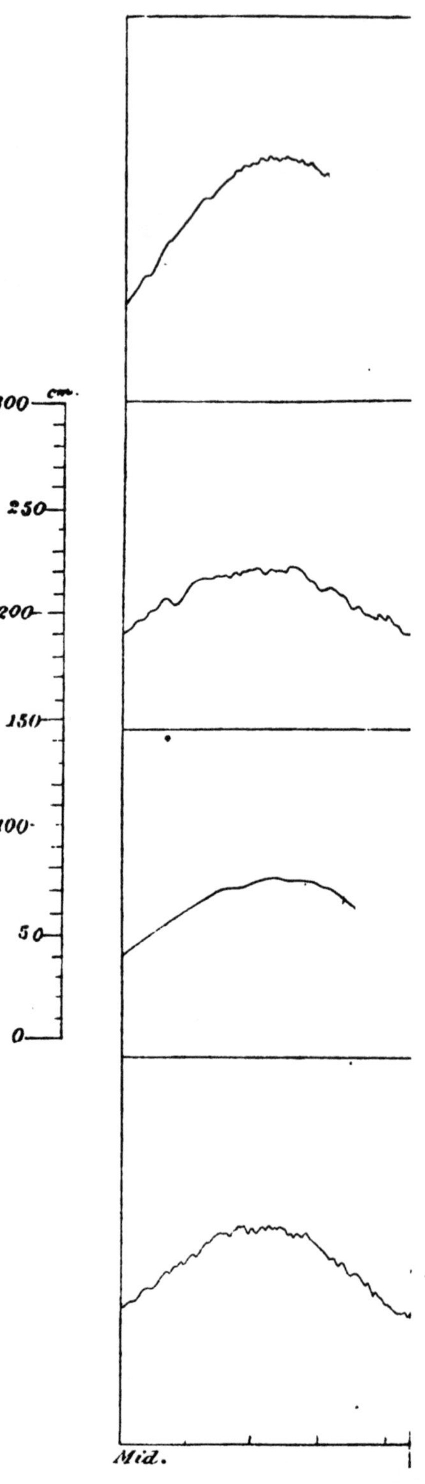

300
cm.
250
200
150
100
50
0
Mid.

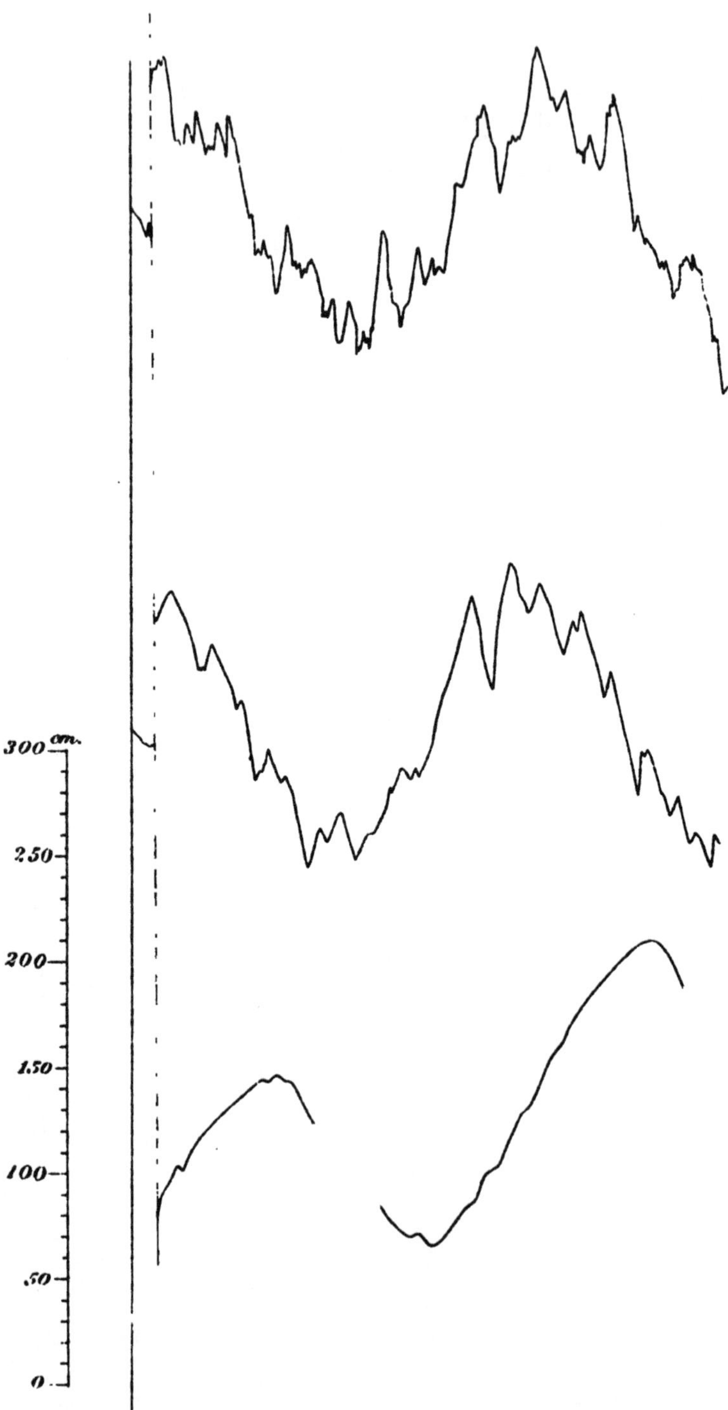
300 cm.
250
200
150
100
50
0

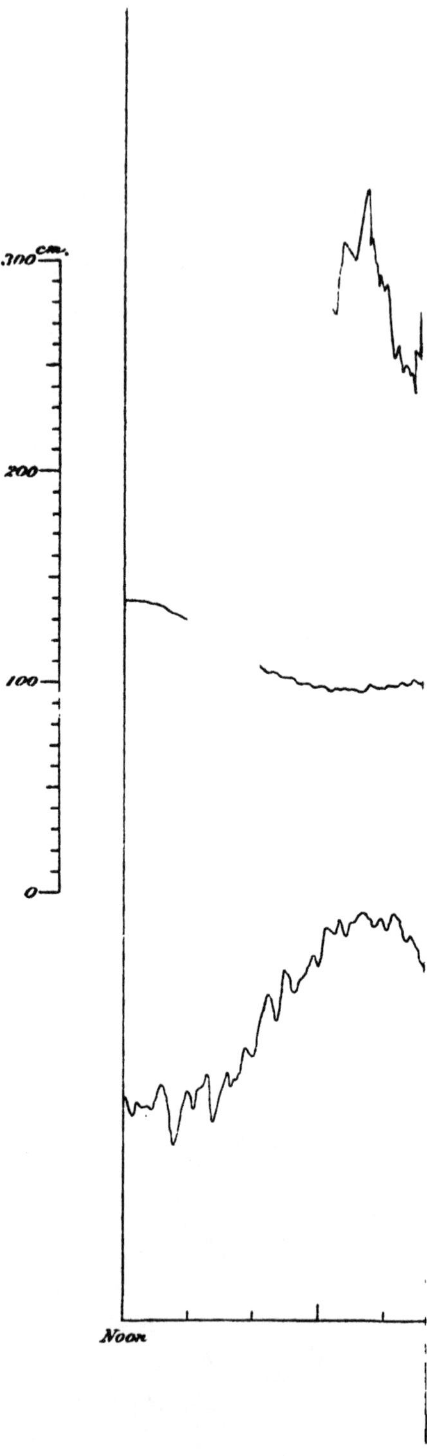
300 cm.
200
100
0
Noon

U (Continued)

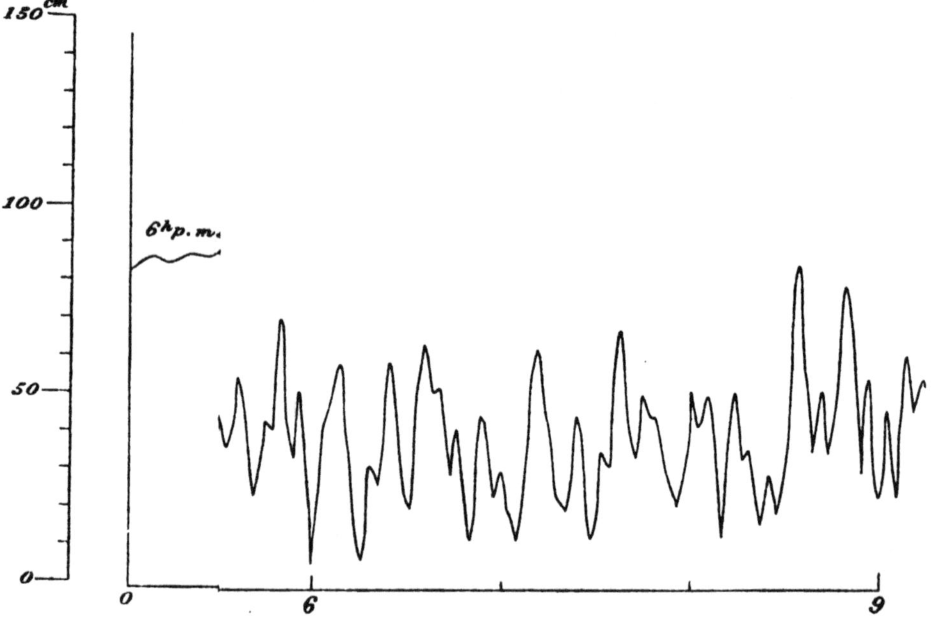

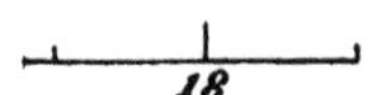

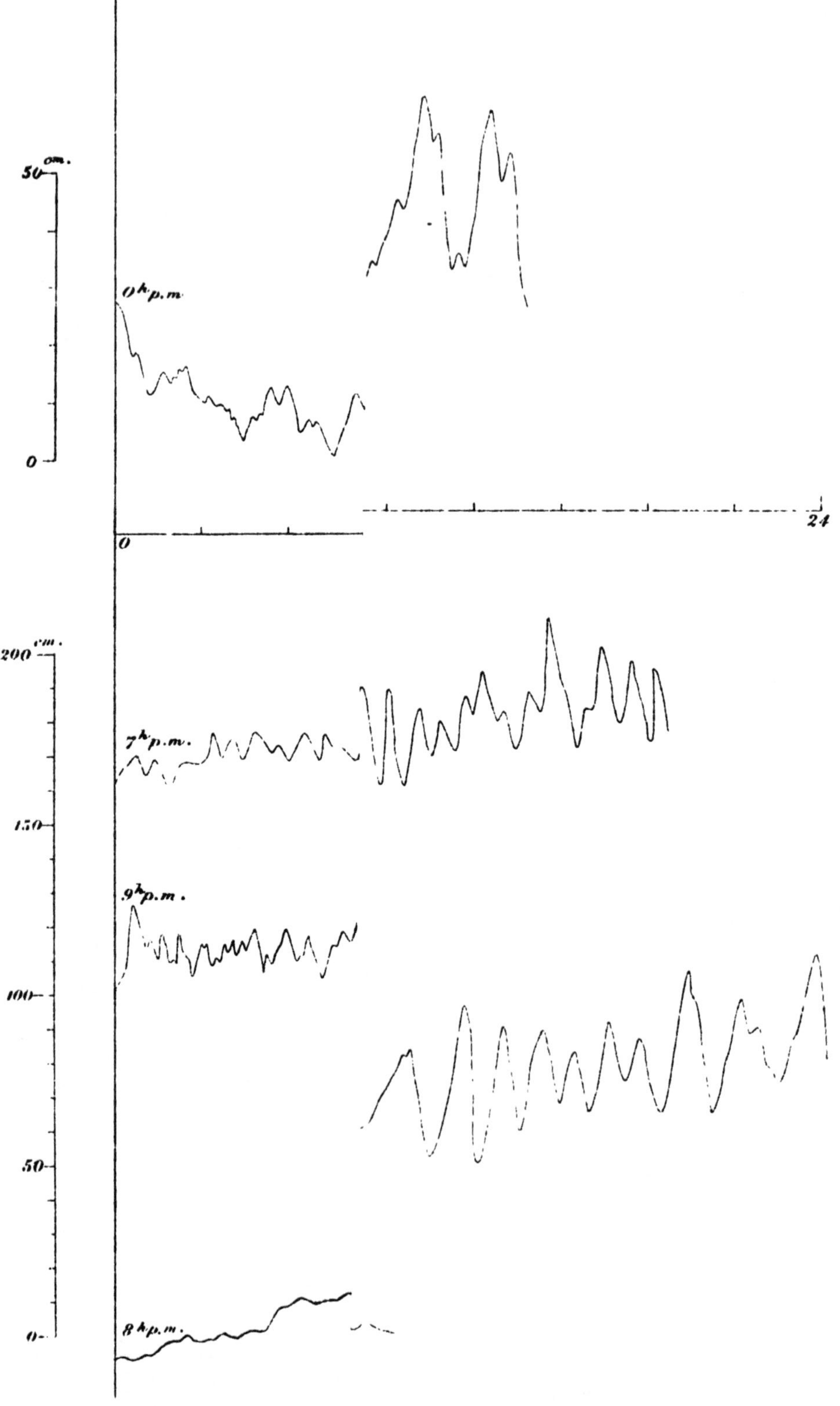
50 cm.
0
0h p.m.
0
24
200 cm.
7h p.m.
150
9h p.m.
100
50
8h p.m.
0

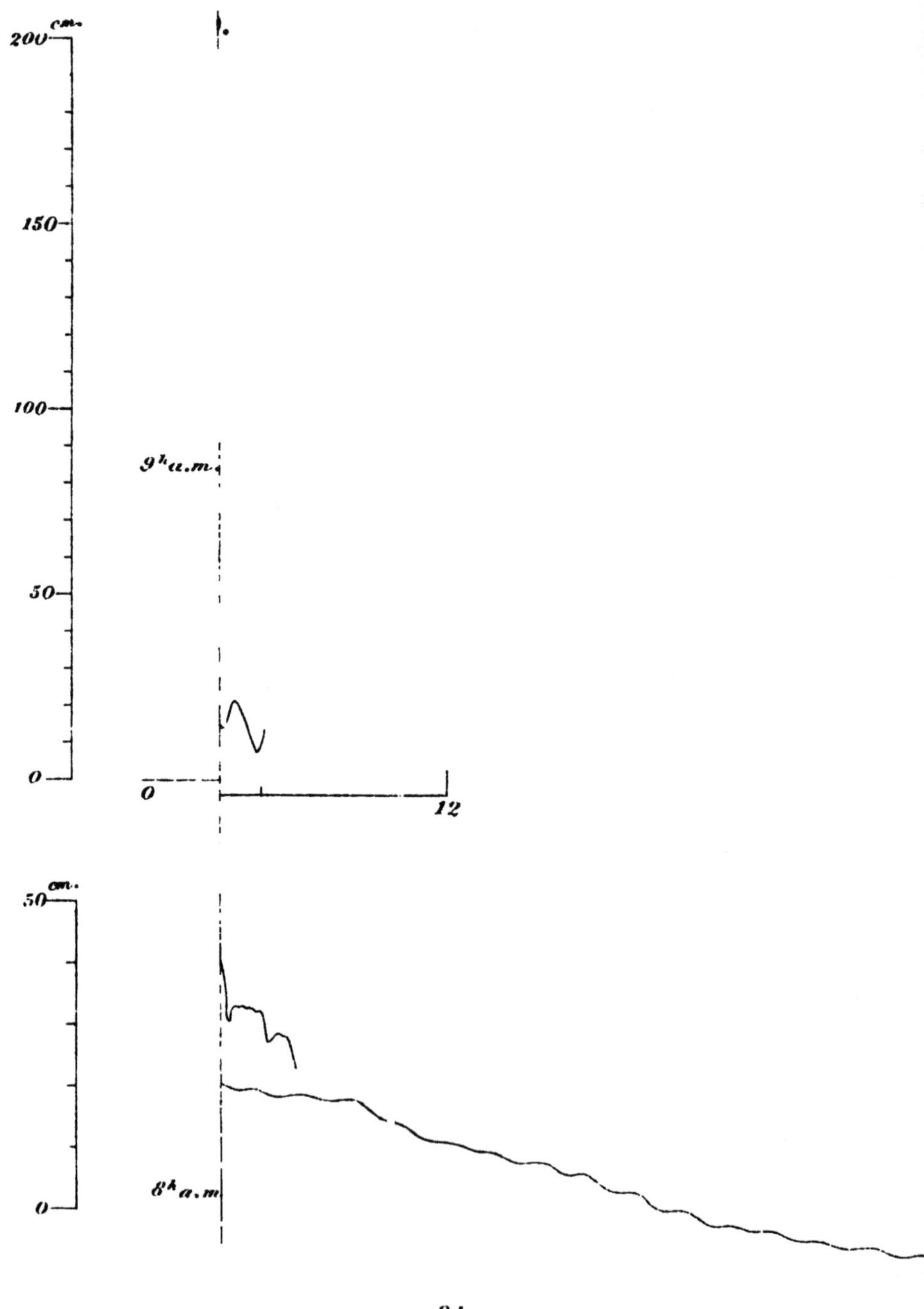

50 cm.

Pl. LIX.

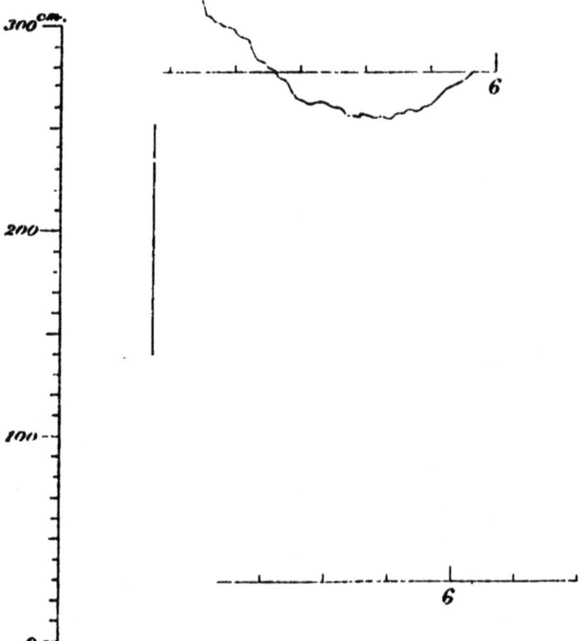

0^h p.m

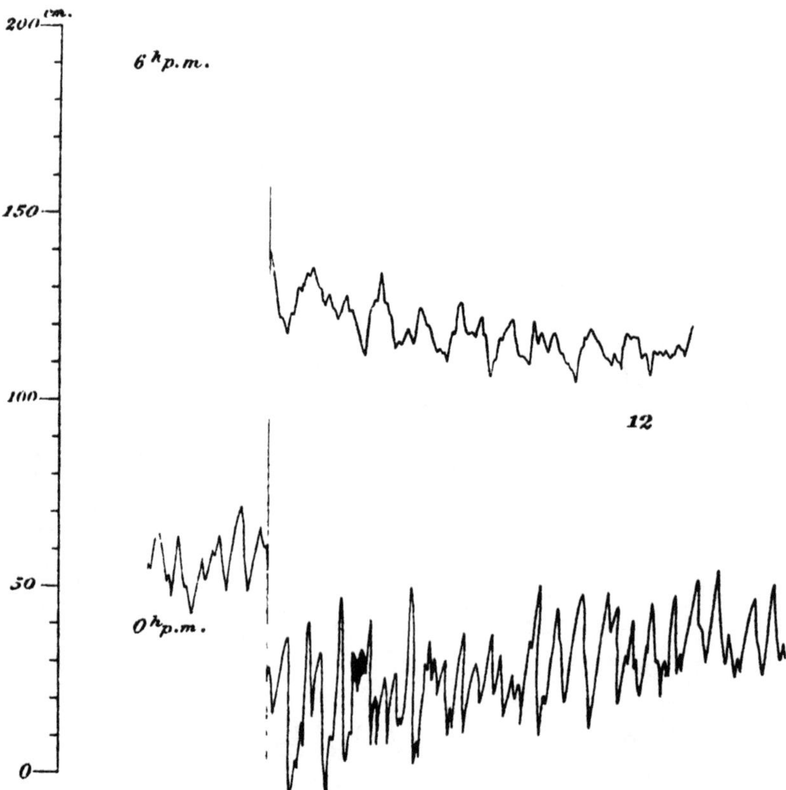

12

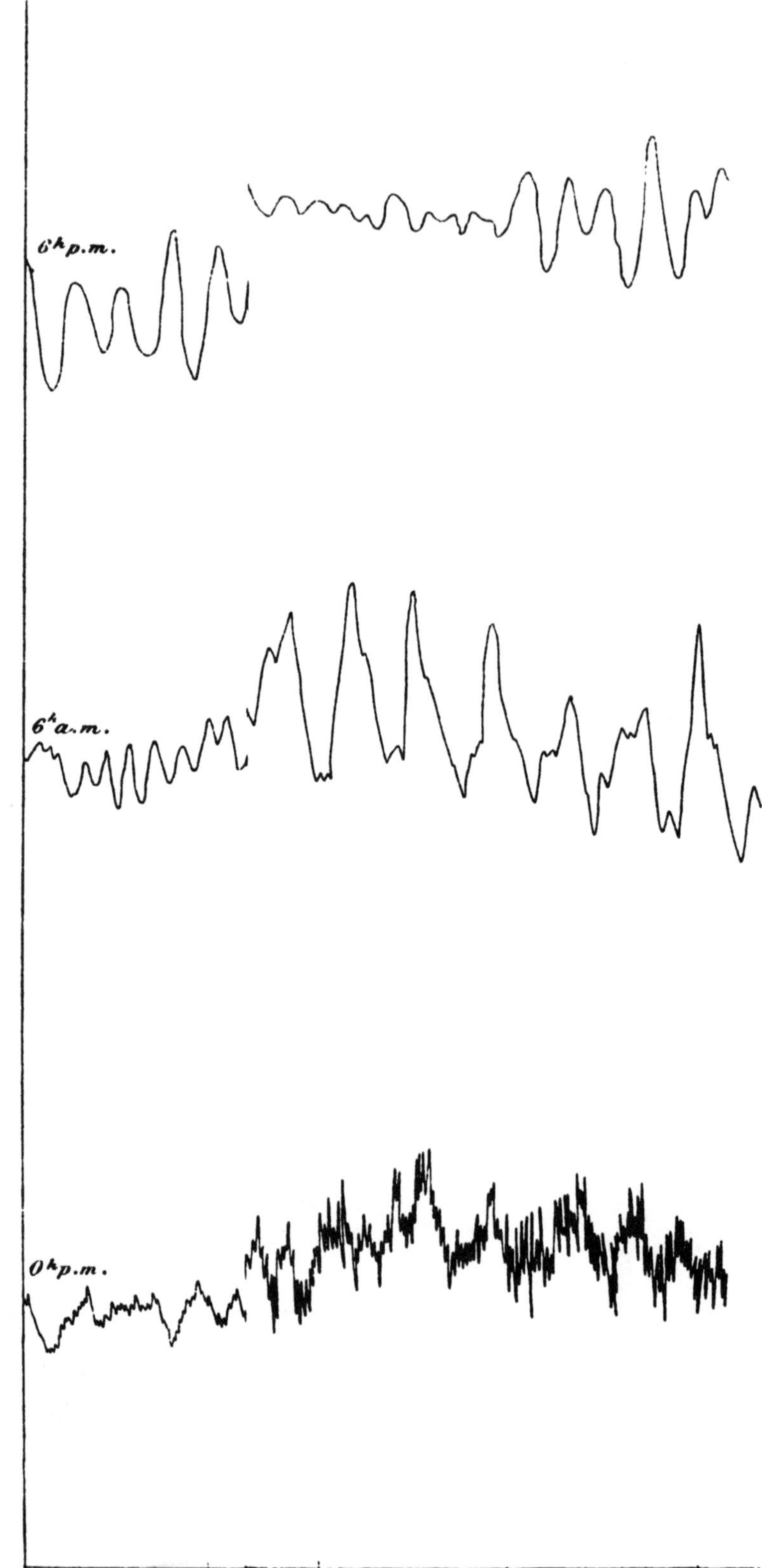
6h p.m.
200 cm.
150
6h a.m.
100
50
0
0h p.m.

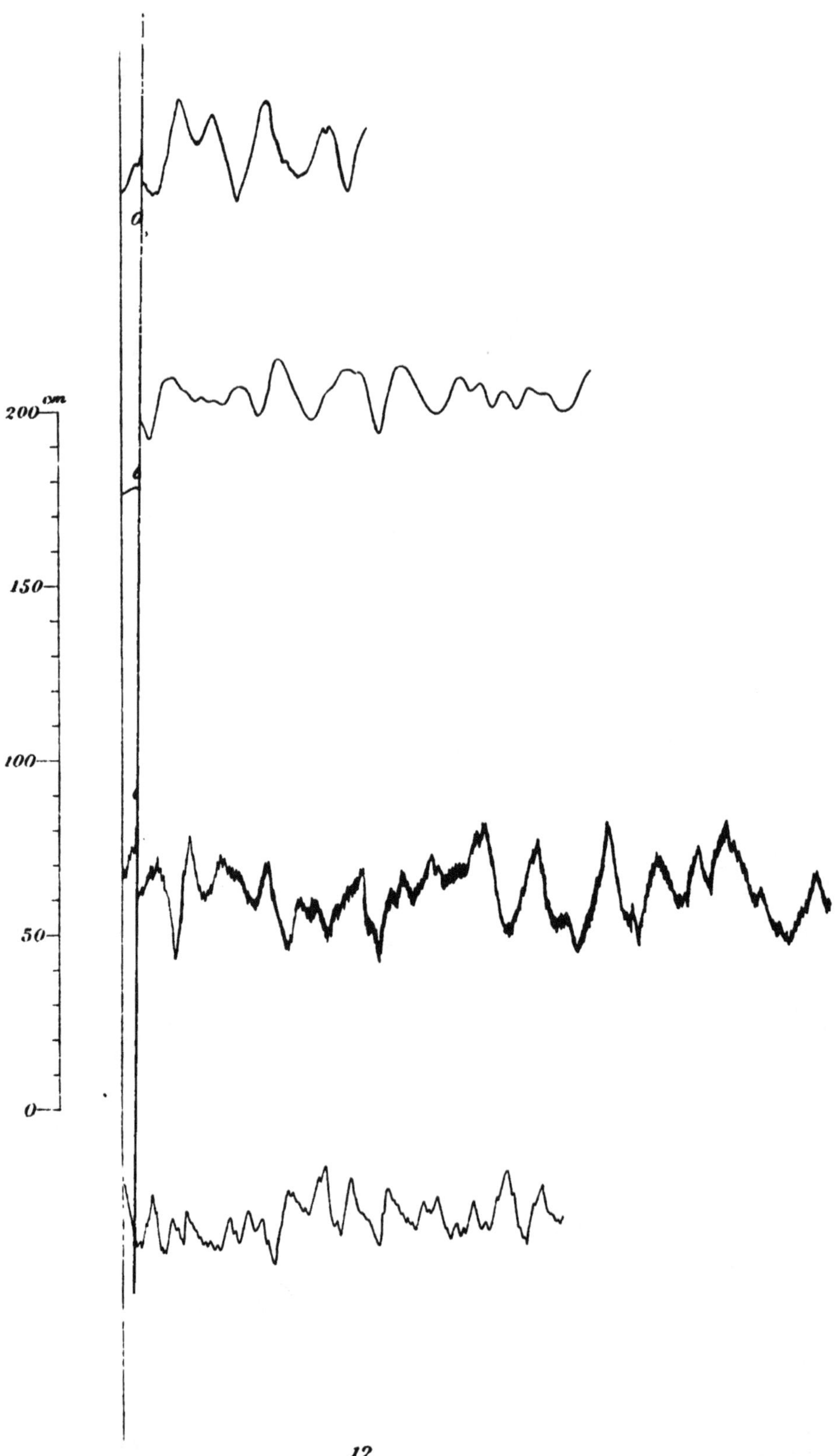

12

The principal tidal components are eliminated in these diagrams by means of the rectifier (see p. 107–109).

Mid.

Aug 31th
Noon

MAPS.

PL. LXV-LXXXVI.

EXPLANATIONS.

Stations actually observed are marked with ○, while those not yet observed are shown by ×.

Soundings are given generally in meters, if not otherwise stated. Coutour lines are given by -----------.

Median lines and mouth lines are shown by —·— — —.

Orientation of each map is shown by an arrow which points to north.

CPSIA information can be obtained
at www.ICGtesting.com
Printed in the USA
BVHW091210191118
533510BV00009B/722/P